SpringerBriefs in Computer Science

For further volumes:
http://www.springer.com/series/10028

Fayçal Bouhafs • Michael Mackay
Madjid Merabti

Communication Challenges and Solutions in the Smart Grid

 Springer

Fayçal Bouhafs
School of Computing and Maths
Liverpool John Moores University
Liverpool
Merseyside
United Kingdom

Madjid Merabti
School of Computing and Maths
Liverpool John Moores University
Liverpool
Merseyside
United Kingdom

Michael Mackay
School of Computing and Maths
Liverpool John Moores University
Liverpool
Merseyside
United Kingdom

ISSN 2191-5768   ISSN 2191-5776 (electronic)
ISBN 978-1-4939-2183-6   ISBN 978-1-4939-2184-3 (eBook)
DOI 10.1007/978-1-4939-2184-3
Springer New York Heidelberg Dordrecht London

Library of Congress Control Number: 2014951918

Printed on acid-free paper

Springer is part of Springer Science+Business Media (www.springer.com)

Preface

The current generation of power networks are struggling to cope with the on-going rise in demand from both commercial and residential users. Initially designed to supply local geographic areas, the electricity grid has evolved over time into a highly interconnected complex system that spans overs hundreds of miles. Today's power grid is comprised of thousands of power lines, power sub-stations, transformers, and other equipment responsible for the energy supply process from fossil-fuelled power plants to the consumer.

Moreover, the centralised and complex nature of the current power grid is creating reliability and scalability issues. These reliability issues, coupled with the need to replace fossil fuel with more accessible and less polluting sources of energy, are driving the modernization initiative of the electricity grid, often called the Smart Grid. The smart grid initiative aims to provide a number of solutions and concepts that promise to change the way we produce, supply, and consume energy. The future power grid will be characterized by novel applications that will help consumers to interact directly with their energy suppliers and involve them in the economics of demand and supply. It will also help consumers to optimise their living style, rearrange their day-to-day energy usage schedule, and in turn enable them to reduce bills from a variety of energy consumption in the house.

Advances in the areas of information and communication technologies, such as wireless communications and wireless sensor networks, will also be fundamental to the smart grid. These technologies are now reaching a level of maturity such that they can be adapted for use in power grid operations. The future power grid will be built on top of on a new pervasive IT infrastructure that provides high level coordination, monitoring, and control capabilities far beyond what is used today. This IT infrastructure will need to provide fast bidirectional communications among all devices and entities involved in electricity production, supply, and consumption. As such, although the requirements of the smart grid will vary from one application to another, communications in this infrastructure will be ubiquitous and must be reliable, flexible and fault tolerant.

This book will discuss the rise of the smart grid from the perspective of computing and communications and aims to explain how current and next-generation network technology and methodologies can help realize the potential that the smart grid initiative promises. In this book, we will study the control and communication issues that need to be addressed in order to modernize the power grid and introduce smart grid applications, identify the requirements that need to be met in order to achieve this objective, and present the technologies that will help in implementing the smart grid.

Contents

Chapter 1
Overview of the Smart Grid

Since its introduction in the early nineteenth century, the electrical grid has played a crucial role in supporting the development of industrialised societies and communities. Its ability to convey both energy and information makes electricity very important to the economy as it enables an array of services, products and applications (such as heating, communication, lighting, and entertainment). The power grid is now largely ubiquitous and connects buildings, homes, as well as factories, campuses, complexes, and hotels.

Initially, power grids started as insular systems to supply local geographic areas. However, these relatively small size local networks evolved over time and became interconnected to achieve more economical and reliable power supply. Today's power grid is very large and highly interconnected on a national scale, compromised of thousands of power stations delivering power to major load centres via high capacity power lines which are then branched and divided to provide power to smaller industrial and domestic users over the entire supply area. The topology of the existing power grid has also been devised around traditional power generation technologies where large power stations fuelled by coal, gas, or oil, have been and remain the main source of electricity generation. These large power stations are built far from consumers in order to be closer to the fuel source, such as coal mines, or to minimise safety concerns as with nuclear power plants. As such, high voltage lines are needed to deliver electricity to urban areas where transformers and complex distribution networks are used to supply electricity consumers, as depicted in Fig. 1.1.

While the topology of the current power grid and its electricity supply strategy has proved efficient and reliable for most of the last century, the introduction of high power consuming applications such as heating and air conditioning results in ever-larger daily peaks in electricity demand. During these peak periods, certain parts of the power grid can become overloaded and fail, causing power outages that could propagate to other parts of the grid, resulting into a partial or total blackout. The current power grid is experiencing equipment failures almost daily and the last decade has witnessed several blackouts in many countries [1, 2]. For example, the

© The Author(s) 2014   1
F. Bouhafs et al., *Communication Challenges and Solutions in the Smart Grid*,
SpringerBriefs in Computer Science, DOI 10.1007/978-1-4939-2184-3_1

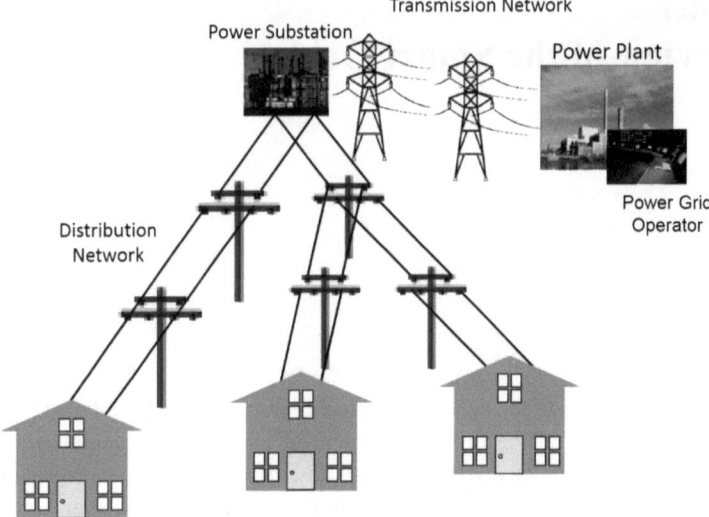

Fig. 1.1 Example of a traditional power grid

blackouts in the Northeast of America in 2003 and in India in 2012 happened during summer when usage of air conditioning systems was at its peak. These blackouts caused disturbance to critical infrastructures, such as transport and water supply, and resulted in significant losses to the economy.

In parallel to these reliability issues, there are growing concerns in some areas such as Europe over the dependence on imported gas and oil necessary to fuel power stations and the environmental damage that these energy resources might cause. Governments in many countries are therefore pushing towards cleaner power production and are encouraging the deployment of green energy resources such as renewables.

These concerns and others are driving the modernization initiative of the power grid that many refer to as the smart grid [3]. This initiative aims to bring the power grid into the twenty-first century by applying digital processing and communications, and introducing computing-based remote control and automation technologies. This chapter will explore the issues with the current power grid in more detail before moving on to introduce the smart grid and the potential benefits it will introduce.

1.1   Smart Grid Drivers

There are a number of technical, environmental, and socioeconomically forces that are driving the modernization of the electricity grid. Current fossil-fuelled power generation is dependent on a limited and expensive resource and are perceived by many governments as strategically unsustainable. In many countries, this technology has been the main source of electricity generation for more than a century, but the centralized nature of power supply dictated by this method of power generation is increasingly creating reliability issues for the power grid. Power outages and blackouts are becoming more frequent causing disturbance to critical infrastructures, losses to economies, and affecting families and communities. These issues, and concerns over the effect of these energy sources on the environment are driving the integration of renewable energy sources into the power grid on a large scale.

1.1.1   Aging Infrastructure

The power grid is one of the oldest infrastructures in many countries and it has been supporting industrialised societies for more than a century. This infrastructure covers wide geographical areas and delivers electricity from large central power generation stations towards consumers via reliable transmission grids. Typically, a power grid consists of power cables that connect power stations to consumers via different transformers and control entities. Although this infrastructure is subject to periodic maintenance and occasional equipment replacement, most of its equipment is old, and struggles to function properly all the time, especially during peak demand periods or in the face of environmental extremes. Understandably, this proneness to failure affects the reliability of the power grid so modernization of this equipment is an ongoing process. However, this modernization process should not be considered merely as the replacement of faulty equipment, over time the need has emerged to incorporate new technology, such as with the introduction of renewable generation, and new equipment and software needs to be designed and installed to reflect this change.

1.1.2   Environmental Impact

It is now widely accepted that greenhouse gas emission, such as carbon dioxide, has a serious and lasting impact on the environment. Due to high emissions of this gas, the so called greenhouse effect is accelerating to the point where it is increasing global temperatures and affecting the climate. Recent studies estimate that fossil-fuelled power generation is responsible for up to 40% of carbon dioxide emission so it is becoming clear that the environmental cost of traditional thermal power plants is therefore unsustainable. Concerns over these environmental effects are driving radical changes in the way we produce, deliver and consume electricity. For instance, the large scale integration of renewable energy sources into the power grid in order to reduce the impact of power generation. Also, on the user side, smart energy management systems are one of the futures technologies considered in the smart grid to help them consume energy more efficiently.

1.1.3   Security of Energy Supply and Increase in Energy Needs

The efficient and reliable transmission and distribution of electricity is fundamental to maintaining functioning economies and societies as we know them. However, most of electricity generation is currently based on power plants that use fossil fuels which is an increasingly limited and contested resource. For example, these natural energy sources are also currently necessary to fuel cars, airplanes, and other transport systems. Fossil fuel is also used to produce fertilizer and pesticides necessary for agriculture and it is central to the production of other materials such as plastics.

As such, the benefits brought by these natural resources to modern societies have created a huge demand on oil and natural gas which is steadily expanding as the world population increases and the economy grows. This dependence of so many economic sectors on fossil fuel and the increasing demand on it has resulted in a rapid depletion of these natural resources so many governments, especially in countries that rely on foreign oil and gas supplies, are now considering the current situation as a threat to the survival of their industrial economies. Moreover, the energy importation process is vulnerable to geopolitical factors such as: international conflicts, political instability in oil producing countries, union strike, etc. As such, many are now taking steps to move critical infrastructures, such as power generation, away from this dependence.

1.1.4   Electricity Cost

The efforts to modernize the electricity grid are also driven by the need to reduce the costs of electricity. Regulators and policy makers are pushing for more affordable electricity tariffs for the consumers but as most electricity generation today is based on fossil fuels, electricity tariffs remain linked to fluctuations in prices on the international energy market. As such, the large scale integration of renewable energy sources into the grid could help break the dependence on oil and gas and reduce the cost of generating electricity. Another factor that is affecting electricity prices is the centralized model of energy supply which inherently comes with an increased cost as the majority of the energy that enters the distribution network is wasted in the form of heat before delivering any useful energy to the consumer. This is in addition to frequent failures caused by old equipment in the distribution network which further increases the cost of electricity supply. As such, distributed generation represents a cheaper and more efficient solution to deliver energy closer to the consumer than the centralized power grid.

1.2   Smart Grid Technologies

The smart grid initiative will therefore be characterized by the integration of a number of technologies that will help improve the reliability and efficiency of electricity supply. These technologies, depicted in Fig. 1.2, will also help to reduce the costs of generating, and supplying electricity to the consumers.

1.2.1   Renewable Energy Generation

Reducing the dependence on fossil fuel for electricity generation is crucial to modernizing the power grid and there are a number of natural phenomena that are being exploited to generate energy, such as: wind, tides, and sunlight. These natural energy sources have the advantage of continually replenishing, hence the energy generated cannot be exhausted, but specialised generators have to be designed to efficiently exploit these resources. For instance, wind turbines are needed to convert airflows into energy; photovoltaic systems are used to convert solar energy into electricity, etc. Moreover, in addition to its reduced environmental impact, renewable energy generation is considered a cheaper option to produce electricity in comparison to oil or natural gas, once one considers the cost of extraction and shipping. In many countries, energy operators are already building power plants that use renewable energy to produce electricity. For instance, there is an increase in the number of wind farms, especially in major industrial countries, with a worldwide capacity of 238,000 MW [4]. Another example of large scale renewable energy production is the Solar Energy Generating Systems in California, which consists of nine solar power plants with a capacity of 354 MW.

Moreover, renewable energy generation is increasing becoming popular among consumers as more and more households install small generators within their prem-

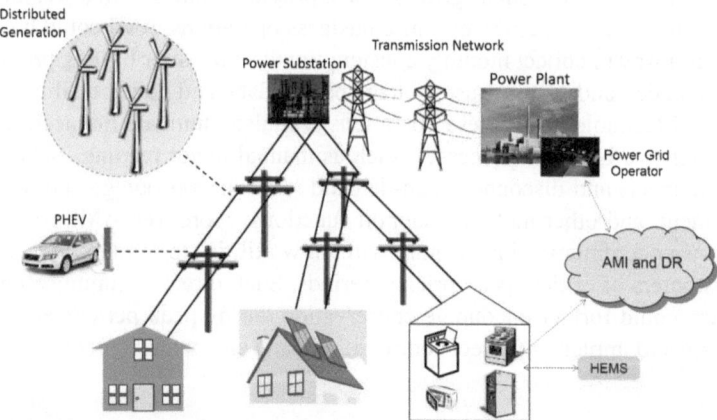

Fig. 1.2 Example of smart grid technologies

ises. Through government subsidies for installation and the potential to sell energy back to the power grid, the aim is to make these systems increasingly economical. Also, due to their proximity to the consumers, these residential energy generators could help to alleviate some of the demand on the power grid during peak times if done on a sufficient scale.

1.2.2   Demand Response

The future power grid will also be characterized by the two way interaction between the consumers and their providers. This feature will enable a range of technologies that will allow collaborative energy management through more ubiquitous monitoring. Demand-Response (DR) [5] is a smart grid technology that will offer utilities the possibility to interact with appliances and electrical devices within customers' homes and buildings and allow them to alleviate the stress on the power grid during peak demand periods by modulating electricity demand. For example, it could provide constant information to the utilities about the energy usage patterns of their customers which will allow them to closely monitor, shift, and balance power load in ways that could optimize energy usage and avoid congesting certain parts of the grid. DR will also be very useful in the context of residential renewable energy generation as discussed in Sect. 2.1, as it will help consumers to trade their energy excess in an energy market.

1.2.3   Advanced Metering Infrastructure

The Advanced Metering Infrastructure (AMI) [6] is another of the main applications in the future smart power grid. AMI will allow utilities to interact with electricity meters enabling real time measurement of energy usage. AMI technology will be a vital component of the smart grid as it will provide utilities with a wealth of new information that could help to optimise business operations. AMI could be used by utilities as a way to collect monthly consumption data used for billing, provide load profile data, demand, time-of-use, voltage profile data, and power quality data. The use of AMI technology for these operations will also eliminate the need for many labour-intensive business processes, such as manual meter reading, field trips for service connects and disconnects, on-demand reads, power outage and restoration management, and other metering support functions. Moreover, AMI systems with this two-way communications feature will allow utilities to send pricing signals to alert customers of critical peak pricing periods. Such direct communication to the customers could further encourage conservation during peak periods and will enable utilities to implement direct control of demand side management.

1.2.4   Home Energy Management

Homes today are equipped with many appliances and electrical devices that consume a significant amount of energy daily. The smart grid initiative aims at providing consumers with tools that could help manage these appliances more efficiency in order to reduce their electricity usage. A major application of the smart grid will be Home Energy Management Systems (HEMS) [7]. This application will enable households to effectively centralise the management of services in a house, provide them with all-round functions for internal information exchange, and help to keep them in contact with the outside world. It will also help households to optimise their living style, rearranging the day-to-day energy usage schedule, and in turn enable them to reduce bills from a variety of energy consumption in the house. This process is starting to be implemented in the form of smart meters that inform the consumer about current energy usage.

1.2.5   Plug-in Hybrid Electric Vehicle

Vehicles are part of our daily lives and represent a major player in world energy consumption and environment pollution. Until recently, research was focused primarily on ways to optimize the design of vehicles and improve their subsystems. However, with the emergence of advanced electronic vehicular technologies, research efforts have been directed towards advancing batteries and storage systems, engine and power train control, and Hybrid Electric Vehicle (HEV) optimization. Plug-in Hybrid Electric Vehicle (PHEV) [8] technology shows great promise as it has the potential to curb emissions and reduce the cost of transportation. Although plug-in vehicles are still not adopted on a large scale, governments, utilities and auto companies are enthusiastically anticipating the opportunities that may arise from reduced emissions and gasoline consumption, new services and increased revenues, and new markets that would create new jobs. PHEV will also have a significant impact on the smart grid as they may result in large spikes in demand as vehicles get charged overnight and represent potential energy generation or storage mechanisms, explored in Vehicle to Grid (V2G) projects.

1.3   Smart Grid Benefits

The modernization of the power grid and the integration of new technologies and applications will help to improve the quality of production and supply of electricity to customers. It will add more intelligence to the control systems governing the power grid and will provide flexibility to its supply operations. Moreover, the smart grid will help to optimise electricity demand on the consumer's side, through technologies such as: HEMS, and advanced metering infrastructures. As such, we can view the smart grid as an end-to-end upgrade of the existing power grid.

1.3.1 Technical Benefits

The applications and technologies of the smart grid will help address many current technical challenges in power supply. Intelligent and decentralised control that replaces traditional centralised control systems will increase grid flexibility and the robustness of its energy supply operations and Advanced monitoring technologies such as wireless sensors and phasor measurement units will improve power failure detection and diagnosis. Consumer-based applications such as smarts meters, and advanced metering infrastructures will help operators to monitor the quality of energy supply, and will allow power loss detection in real time. This will in turn also speed up the power restoration process, and minimise the impact of power outages. Demand-Response systems will help reduce the stress on the power grid during peak demand periods, thus, avoiding congestion and power failures. When associated with distributed renewable energy generators and storage units, these systems could improve the quality of supply and relieve parts of the grid that are under much of the stress.

1.3.2 Economic Benefits

Improving power grid reliability and addressing the technical issues that face them, will also bring significant economic benefits to the consumers. Through novel applications and technologies, the smart grid will contribute to reducing the frequency of power outages and equipment failures which will result in less maintenance operations and lower the overall cost of electricity supply. The integration of renewable energy sources into the grid will also reduce the dependence on imported oil and natural gas, and their impact as a result of price fluctuations in international markets.

Moreover, the improved reliability of the grid will help to increase the efficiency of operations in industries and businesses. Any power outrage affects the production and productivity of the industries and businesses that rely on continuous electricity supply for their daily operations, with the losses in productivity often passed on in the form of higher prices for their products and services. Therefore, a reliable power grid will contribute in reducing the cost of production and productivity.

Households will also reap a significant economic benefit from the smart grid and its applications, especially smart energy management systems and smart metering. These applications will enable consumers to monitor their electricity usage more accurately, and help them to make better decision regarding the utilization of their appliances.

1.3.3 Environmental Benefits

As we have seen, due to their dependence on fossil fuels power grids are a major source of environmental pollution. The modernization of the power grid and the optimization of the electricity supply process will reduce CO_2 emissions through distribution and the introduction of renewable energy sources will break the dependence on fossil fuel to generate electricity, and further reduce pollution caused by production. On the demand side, the smart energy management systems within homes, smart meters, and AMI will help consumers reduce their electricity usage which will further contribute to achieving this objective. Finally, PHEV Vehicles will also bring benefits to the environmental as this technology starts to replace internal combustion vehicles.

1.4   ICT in the Smart Grid

Information availability is very important for many smart grid applications, and therefore Information Technologies will play a major role in the smart grid initiative. Many of these technologies, such as wireless communications and wireless sensor networks, are already being used in other sectors such as manufacturing and telecommunications. These technologies are now reaching a level of maturity such that they can be adapted for use in power grid operations.

1.4.1   Digital Communications

Digital communications are paramount to control systems in the current power grid and will become more so in the smart grid. The power grid is typically monitored by *Supervisory Control And Data Acquisition* (SCADA) [9] systems that collect data from control devices located at strategic points within the power network. However, the legacy and proprietary nature of these control systems is reflected on the underlying communication networks, being typically low bandwidth and inflexible. In the smart grid, this legacy communication infrastructure will be replaced by uniform and standardized communication technologies as we see in internetworking today. The communications infrastructure will need to be extended and its bandwidth capacity enhanced in order to support the new range of applications such as AMI, and Energy Management System as well as the ubiquitous monitoring and sensing devices required. Through its protocols and services, the future communication infrastructure will provide flexibility, reliability, and redundancy to support these features.

1.4.2   Wireless Sensing

Current control operations in the power grid are based on measurements taken from key points within the distribution infrastructure. These measurements help operators monitor the health of equipment and power lines, supervise the quality of electricity supply, and devise optimal management strategies. Power grid operators rely on field sensors to obtain these measurements and most of these sensors today use a wired communication medium, such as telephone lines, to send their readings to the operators' control machines. However, this increases the cost of deployment and limits the number of sensor that can be deployed into the power grid due to scalability. In addition, these sensors suffer from the same legacy and proprietary nature that characterize current control systems.

Future smart grid applications will require far more measurements from the grid and so sensors will need to be deployed more densely, and need to take measurements more frequently. The latest developments in wireless communication technology and micro-electro-mechanical systems (MEMS) have enabled the development of new kinds of wireless sensing technology which come integrated with sensing, data processing and communicating components, and can communicate untethered over short distances [10, 11]. In addition to their low cost and easy deployment, the wireless nature of these devices will facilitate the gathering of the fine-grained measurements required by smart grid applications. Wireless sensing technology will also be very important for home energy management applications to gather information such as indoor temperature and people's movements, that could further help optimize energy usage in homes.

1.4.3   Distributed Systems

In addition to fast communication and sensing capabilities, the realization of the smart grid will require a high performance management infrastructure capable of providing dynamic intelligent control decisions. This control infrastructure will need to provide global analysis of all events observed in the power grid in order to respond rapidly to any adverse situation. Today's power grids are managed by centralized control systems that were designed and implemented decades ago and these centralized systems lack the scalability and flexibility required. As such, they are often slow to respond to emergency situations. The smart grid will therefore represent a major shift from this centralized control approach towards distributed frameworks that use local and coordinated control points. In this model, control operations will be pushed from the utilities control centre towards intelligent control units located at key points within the grid. The control software will be implemented using software technologies such as multi-agent systems [12, 13].

1.5   Summary

The smart grid is the vision to modernize the end-to-end power grid through a number of technologies and applications. Driven by political, environmental, and socioeconomic factors, the smart grid promises to change the way we produce, deliver, and consume electricity. Bulky and polluting power stations that rely of fossil fuel to produce energy will progressively be supplemented and replaced by cleaner renewable energy resources. Technologies such as Advanced Metering Infrastructures and smart meters will allow real time interaction between utilities and their consumers and help optimize their electricity supply operations. These technologies will ultimately help the consumer to manage their energy usage more efficiently and make savings on their electricity bills.

Information and digital communication technologies will also play a major role in realizing the smart grid, as it will be built on a new IT infrastructure that provides high level coordination, monitoring, and control capabilities beyond what is used today. This IT infrastructure will need to provide fast bidirectional communications among all devices and entities involved in electricity production, supply, and consumption. As such, although the requirements of the smart grid will vary from one application to another, communications in this infrastructure will be ubiquitous and must be reliable, flexible and fault tolerant.

In the next chapters we will describe in detail how these technologies will help in implementing the smart grid and we will identify the requirements that need to be met in order to achieve this objective. We will also review the major progress made in the area of communication and distributed systems that could facilitate this modernization process. Chapter 2 will study the control and communication issues that need to be addressed in order to enable the introduction of distributed generation into the power grid at large scale. Chapter 3 will review the smart grid technologies that allow utilities to interact with their customers and monitor their energy usage more efficiently, and identify the communication requirements of these technologies. Chapter 4 will review existing communication technologies that could be used to address the issues identified in the two previous chapters. In Chap. 5, we analyse the communication requirements of home energy management system, and in Chap. 6 we review communication technologies that could be used to realise these systems. Finally, in Chap. 7 we review the data processing and storage challenges in the smart grid, and the approaches that could be used to address this challenge, and present our overall vision for the end-to-end smart grid.

References

1. I. Dobson, *Cascading Network Failure in Power Grid Blackouts, in Encyclopedia of Systems and Control*, pp. 1–5, Springer, 2014.
2. H. Glavitsch, *Large-Scale Electricity Transmission Grids: Lessons Learned from the European Electricity Blackouts*. Wiley Handbook of Science and Technology for Homeland Security, 2014.
3. H. Farhangi, *The Path of the Smart Grid*, in IE, Power and Energy Magazine, vol. 8, n. 1, pp. 18–28, 2010.
4. *Global Wind Energy Council (GWEC)*: http://www.gwec.net/
5. S. Choi, et al. *A Microgrid Energy Management System for Inducing Optimal Demand Response*. in 2011 IEEE International Conference on Smart Grid Communications (SmartGridComm), Brussels, Belgium.

6. D.G Hart, *Using AMI to realize the Smart Grid*. in IEEE Power and Energy Society General Meeting-Conversion and Delivery of Electrical Energy in the 21st Century, 2008.

7. N. Javaid, et al. *A Survey of Home Energy Management Systems in Future Smart Grid Communications*. in Eighth International Conference on IEEE Broadband and Wireless Computing, Communication and Applications (BWCCA), 2013.

8. C. Mi, M.A. Masrur, and D.W. Gao, *Plug-in Hybrid Electric Vehicles*, in Hybrid Electric Vehicles: Principles and Applications with Practical Perspectives, p. 107–138, John Wiley and Sons, 2011.

9. A.A Boyer, *SCADA: Supervisory Control and Data Acquisition*. 4th Edition, 2009, International Society of Automation.

10. V.C Gungor, B. Lu, and G.P. Hancke, *Opportunities and Challenges Of Wireless Sensor Networks in Smart Grid*, in IEEE Transactions on Industrial Electronics, vol. 57, n. 10, pp. 3557–3564, 2010.

11. V.C Gungor, et al., *A Survey on Smart Grid Potential Applications and Communication Requirementsm*, in IEEE Transactions on Industrial Informatics, vol. 9, n. 1, p. 28–42, 2013.

12. M. Schumacher, *Objective Coordination in Multi-Agent System Engineering: Design and Implementation*, in Lecture Notes in Computer Science, Springer, pp. 9–32, 2001.

13. E.M. Davidson,. and S. McArthur, *Exploiting Multi-agent System Technology within an Autonomous Regional Active Network Management System*. in International Conference on Intelligent Systems Applications to Power Systems, 2007.

Chapter 2
Communication for Control in Heterogeneous Power Supply

The need to modernize the power grid infrastructure, and governments' commitment for a cleaner environment, is driving the move towards clean renewable sources of energy and many countries are now building power plants using green energy sources such as wind turbines and photovoltaic systems. Whilst these big energy generation stations will certainly help in addressing the environmental and energy dependence aspects, the related power grid management and congestion avoidance issues during peak demand remains a serious challenge that needs to be addressed.

One way of solving this problem is to bring these renewable energy generation sources closer to the consumer by connecting them directly to the distribution network. In many countries, consumers are encouraged by their respective governments to install renewable energy generators on their premises. In many cases these residential energy generators are capable of covering the consumer's electricity usage and could represent a cheaper alternative to electricity supplied from the central power station in the long term. In particular, the supply of isolated locations with electricity comes at a greatly increased cost to the utilities as the majority of the energy that enters the system is wasted in the form of heat before delivering any useful energy to the consumer. Therefore, bringing the electricity source to the consumers might be a more economical and reliable option.

In the smart grid, it is envisaged that these so called Distributed Energy Resources (DER) will become increasing pervasive within the distribution network. Moreover, DER will not be limited to satisfying local users' electricity demands, but could also use any excess to supply neighbours, and local communities. Ultimately, incentives offered by a future deregulated energy market will help consumers to choose the source of their electricity in near real time. For instance, a consumer could alternate between utilities electricity, and electricity provided by neighboring residents according to the price of electricity at a given time.

The implementation of this heterogeneous energy supply paradigm, however, implies a radical change from the centralized one-way electricity supply that characterizes the current power grid, to a two way electricity supply paradigm. Flexibility will be paramount to realize this vision, and therefore the system that governs control operation in the grid will need to be redesigned, particularly with regard to

© The Author(s) 2014

F. Bouhafs et al., *Communication Challenges and Solutions in the Smart Grid,*
SpringerBriefs in Computer Science, DOI 10.1007/978-1-4939-2184-3_2

the communications aspects. In current control systems, these facilities are generally monitored by a relatively limited number of sensors that are deployed at critical places in the distribution network. This sparse deployment of control devices limits the penetration of the control system and does not provide an extensive, accurate real-time view of the network's status. Therefore, these control systems will not be able to address the challenge of ensuring a stable and healthy distribution network incremented with distributed generation as envisioned in the smart grid. In this chapter we will emphasize the importance of communications and wireless technology in modernizing control systems to support electricity flow management in the smart grid. We will present the limitations of existing control systems and the communications challenges that need to be addressed in order to provide more reliable control systems.

2.1 Control in Traditional Power Networks

The power network covers wide geographical areas and delivers electricity from large central power generation stations towards consumers via reliable transmission grids. Typically, a power grid consists of power cables that connect the power station to the consumers and different transformers and control entities, as depicted in Fig. 2.1. Transmission networks are used to send high voltage electricity towards urban areas, where distribution networks are then used to supply medium and low voltage electricity to consumers. These networks consist of substations, transformers, poles, and load wires that connect consumers to the power grid.

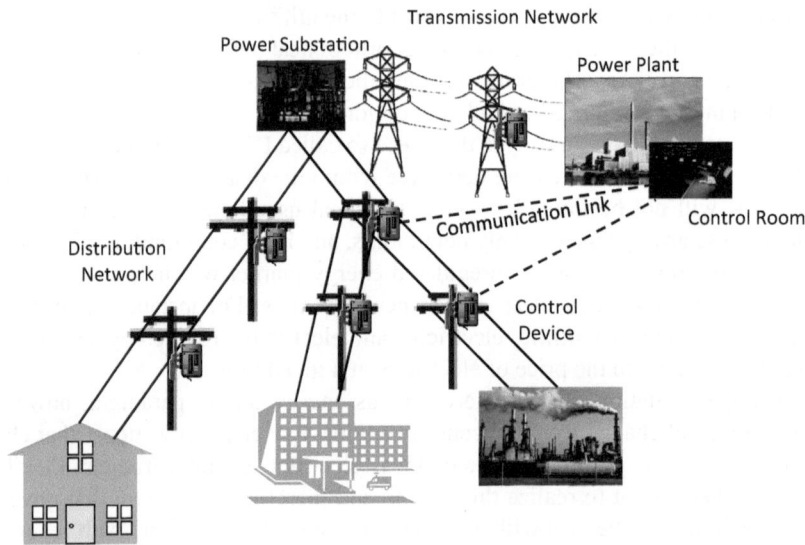

Fig. 2.1 Example of a power grid and its components

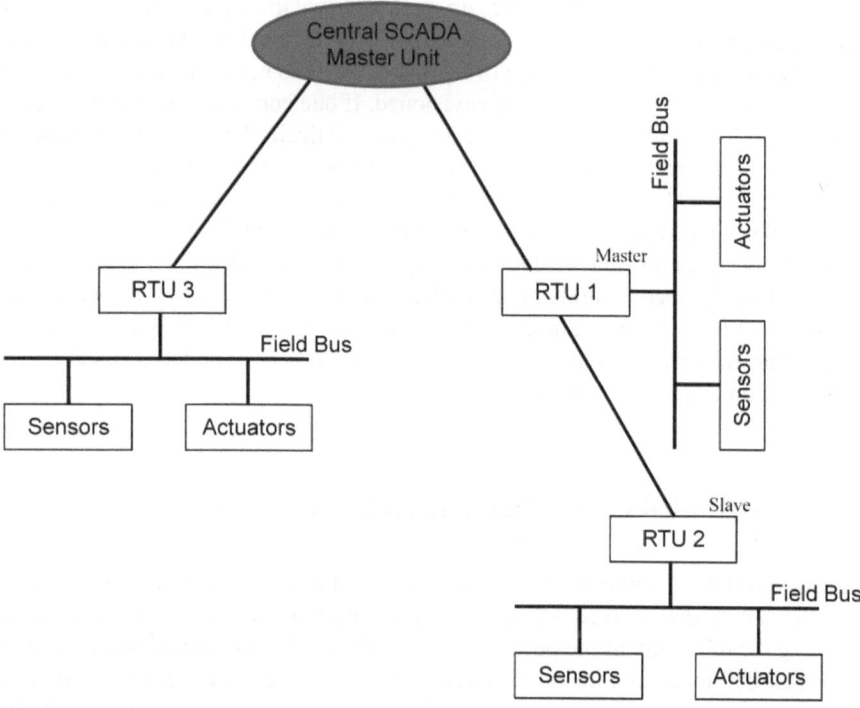

Fig. 2.2 Remote terminal unit in SCADA systems

Currently, power grids are monitored by legacy Supervisory Control And Data Acquisition (SCADA) [1] systems that are composed of control devices, generally called remote terminal units (RTUs), as depicted in Fig. 2.2. The RTU represents a contact point with field sensors and actuators through a field bus. It allows the SCADA system to collect measurements from sensors such as current sensors, and to send control commands to actuators, such as relay points and circuit breakers. In addition, the SCADA system comprises a central Master Unit computer, which is usually composed of one or more servers that represents the interface between the operator and the SCADA system. The role of the central host computer is primarily the processing of data collected from different field-based control devices and presenting them in a readable format to the operator. The communication network is central to this process as it allows data transfer between the control center and the field-based devices, and it comprises typical IT equipment such as routers, switches, modems, etc. These legacy systems were built around a hierarchical and centralized communication approach, where an RTU sends its messages to its direct one-hop RTU neighbor that acts as its master, usually using a multi-drop communication mode. This process is repeated until the data reaches the master processing unit of the SCADA system.

However, as the power grid scales up, particularly with regard to the number and type of monitoring devices being deployed, it is clear that this system may run into a number of issues. First, this type of SCADA system is not designed to handle the number of connected devices being envisioned. If one considers that the number of devices deployed into such a network might realistically be expected to grow by several orders of magnitude then it is easy to appreciate the problem of aggregating data at master RTUs close to the Master Unit. Moreover, such a monitoring network will need to support the much richer information supplied by modern devices, necessitating the provision of dedicated high bandwidth links. Finally, as we have seen, failures in such equipment are entirely possible and hence and system must be designed to be robust. Current systems are not designed in this way and new approaches may well look to adopt the lessons learnt from Internet engineering for managing the communication infrastructure.

2.2   Distributed Generation and Active Control

The integration of small renewable generators into power networks will offer a cheaper supply option to consumers and help supply isolated consumers, hence increasing the reliability of the power grid and reducing its operation costs. However, it is clear that managing this new system will also require a new control system to effectively manage it. This new system will require utilities to move away from the traditional centralized approach and adopt a more distributed, collaborative, and dynamic paradigm.

The introduction of DER into the power grid will create new issues for the distribution network operators (DNOs) [2, 3]. While in traditional power grids the electricity usually flows from the central power stations to the consumers, in a smart power grid incremented with DER, the electricity now flows in two directions, either from the station to the consumer or from the consumer DER back into the grid, as shown in Fig. 2.3. To introduce DER, DNOs will therefore be faced with the challenge of making their power distribution networks more flexible and dynamic. For example, the loads in distribution lines have strict thermal and voltage capacities, which cannot be exceeded. As such, the connection of small generators with variable output rates, such as wind turbines, will represent a risk to the stability of the network unless it is managed carefully.

Despite this challenge, this new paradigm offers some very powerful features that could be used to improve grid performance and reliability. Figure 2.4 depicts an example of an area isolated and cut from the power supply due to a fault that occurred on a feeder which resulted in the opening of the breaker B1. Under this new model, the isolated area could be re-supplied with electricity through generator G1, once breaker B2 is closed. In this case, DER represents a cheaper and more robust solution to deliver energy closer to the consumer than with only the centralized DNO power grid.

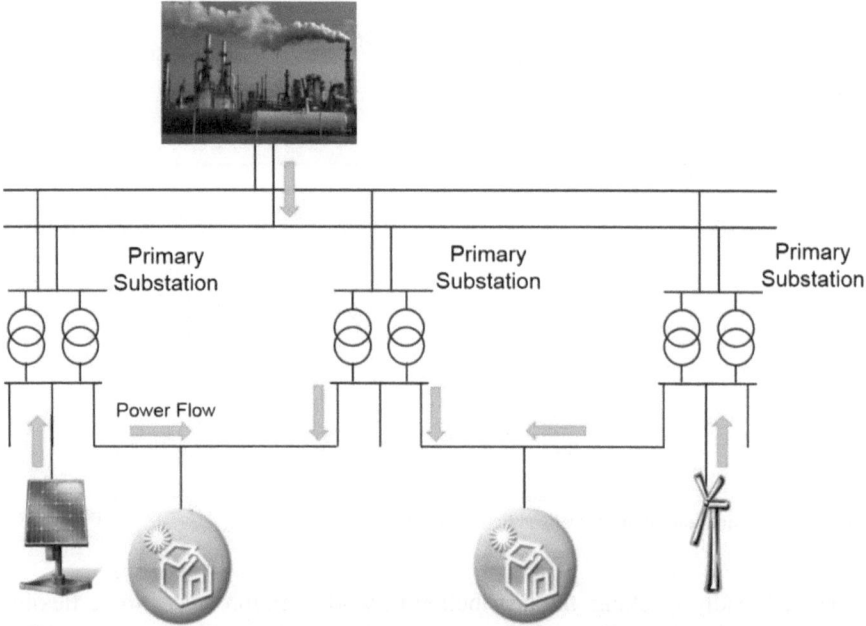

Fig. 2.3 Power grid with distributed generation

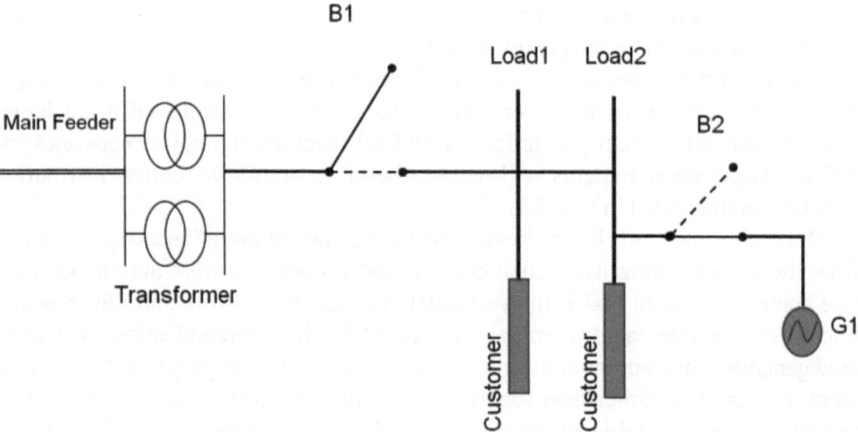

Fig. 2.4 Power re-supply with distributed generation

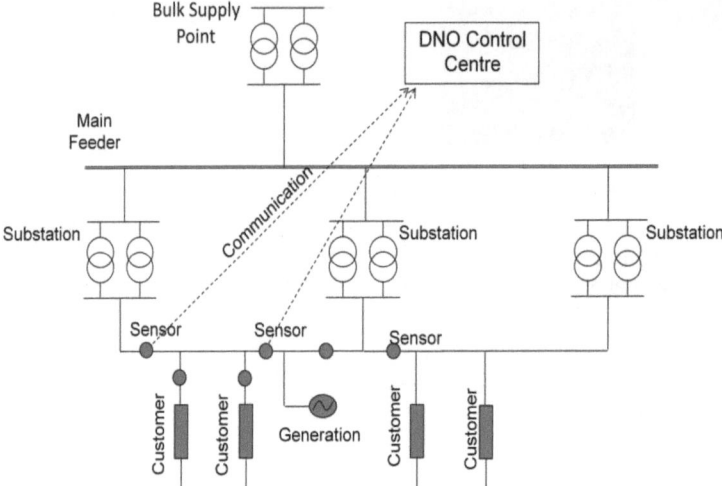

Fig. 2.5 Communication in centralised active control system

More broadly speaking, the distribution network therefore needs to be flexible enough to redirect the flow of excess electricity to other grid segments, or find alternative paths to supply the designated consumers [4, 5]. In such a case, the actual topology of the network will require modification and DNOs' control systems will therefore need to move from passive control to a more active control model whereby the distribution network can be modified and re-configured dynamically according to changes in the power flow [6].

Active control is predicated on continuous real time monitoring and management of the power network. Therefore, sensors need to be deployed in far larger numbers than are currently in order to efficiently monitor the power network conditions. These measurements will need to be taken across the entire distribution network, as illustrated in Fig. 2.5.

Moreover, many works in the area of active control are advocating the move from the current centralised active control model towards a more autonomous active control paradigm [7–9]. In this model, distribution networks are divided into micro-grids, where each micro-grid is formed by the interconnection of distributed generators and autonomous intelligent controllers that are deployed to manage them. In this case, rather than report to the central controller, sensor data will be transmitted to the intelligent controllers and local control decisions will be taken there, as illustrated in Fig. 2.6. In this situation, intelligent controllers will also be required to work in collaboration with neighboring micro-grids and will need to exchange control data [10–12]. Therefore, the availability of a high performance and pervasive communication network is crucial for the realisation of an active distributed control system design.

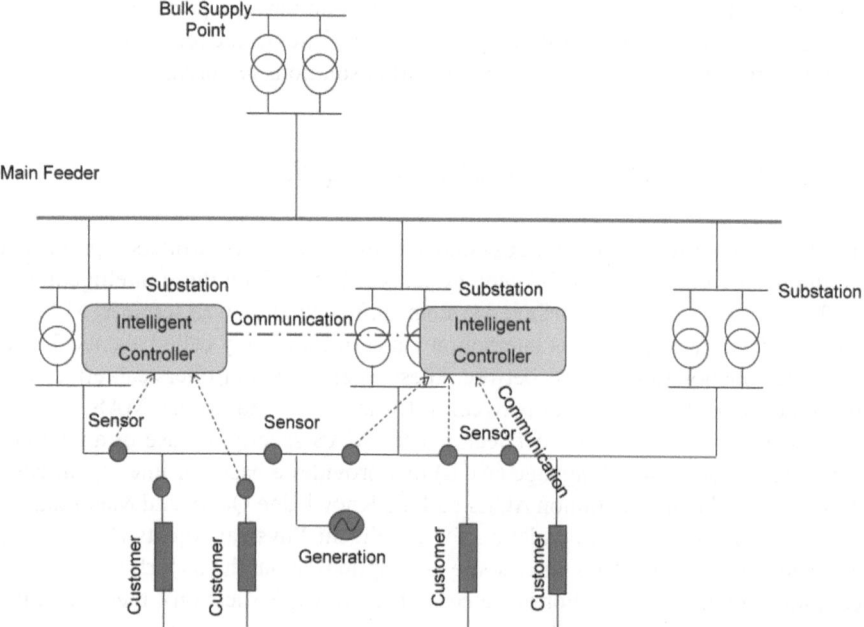

Fig. 2.6 Communication in autonomous active control

2.3   Communications Challenges in Active Control

The realization of a heterogonous power grid that could support distributed renew-able generators in addition to traditional bulk power supply therefore requires flex-ible and active control mechanisms. However, to implement these mechanisms it is important to first understand their communication requirements. The study of active control presented above helps to identify two connectivity requirements:

- *Connectivity between intelligent controllers*: Intelligent controllers need to be interconnected in order to allow for collaborative and distributed control over the power network via micro-grids. This connectivity will allow the controllers to exchange coordination messages before execution of any control action over the power network.
- *Connectivity between intelligent controllers and field-devices*: In order to ex-ecute control actions, controllers both need input from sensors in the power net-work such as measurements, alarms, etc., in addition to the ability to send control commands to actuators in the field.

These requirements show that interconnections will be needed to carry two specific types of data traffic: one will be used to carry inter-controller coordination messages, the other will be used to transmit field-devices monitoring reports. The design of each connection relies on the study of the characteristics of the relevant traffic. This section will help to identify the communication solutions required.

2.3.1   Inter-Controller Coordination Traffic

The implementation of autonomous control with intelligent controllers is predicated on the deployment of a Multi-Agent Systems (MAS) within these intelligent controllers [13, 14] to form a distributed system, akin to modern web services. MAS are characterized by a continuous interaction via entities, usually called agents. Therefore, the collaboration process between these intelligent controllers can effectively be modelled as data exchanges between software agents that form a MAS.

To ensure interoperability between agents, MAS specify the use of a common Agent Communication Language (ACL) that provides a basis for inter-agent communication. The most common ACLs include Knowledge Query and Manipulation Language (KQML) and Foundation for Intelligent Physical Agents (FIPA) [15]. The specifications of ACLs make some assumptions about the underlying transport communication protocols. For instance, in the FIPA specifications, there is an implicit assumption that the underlying data communication network used to transfer information between remote agent platforms is based on the TCP/IP protocol stack, as illustrated in Fig. 2.7. This allows agents to communicate over arbitrary distances with a reasonable degree of reliability.

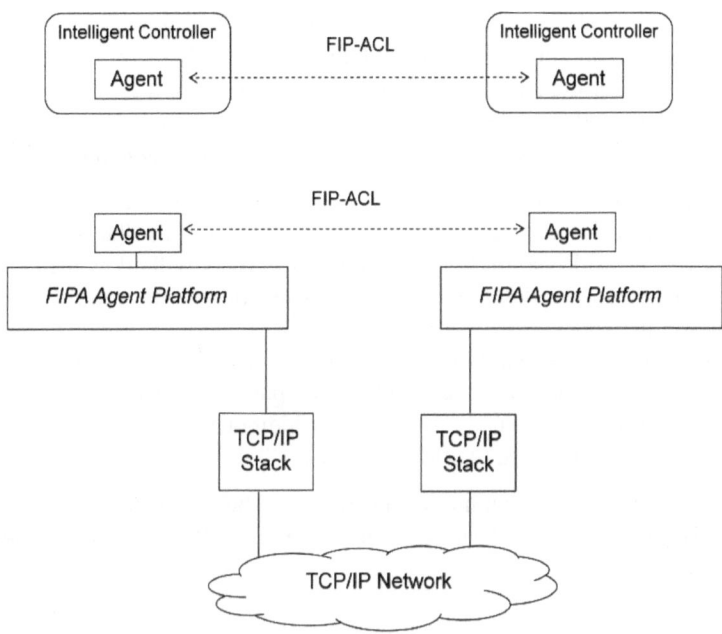

Fig. 2.7 FIPA-ACL communications over TCP/IP

A study published in [15] shows that a simple Request/Agree exchange between two agents based on FIPA-ACL over a communication network generates two messages of sizes: 1408 bytes, and 2854 bytes respectively. Moreover, since the ACL is expected to run over a standard protocol stack, such as TCP/IP, the ACL messages are embedded (encapsulated) into a new message with extra control overhead added at each layer of the protocol stack.

2.3.2   Field-Device Control Traffic

Dense deployment of monitoring devices in the distribution power network is vital to the success of active control. Such dense deployment of control devices, however; will induct extra data traffic that includes all types of measurements and control messages, such as: fault passage indications and reports generated by RTUs deployed in the distribution network. Currently, most of these field devices are using proprietary communication protocols, such as WISP+ and Ferranti [16], to send their data however; in the future these devices are expected to use a common communication protocol, which will facilitate interoperability between devices from different manufacturers.

This data traffic includes all measurement, fault passage indications, and reports generated by RTUs deployed at medium and low voltage distribution network (11 kV). The most important issue with this traffic is therefore to have a unified data model for all the RTUs. Such a unified model will make the communication between the RTUs and the intelligent controller easier, and avoid the need for translation gateways at each end of the communication link. There are many standards that have been proposed in this area that aim to provide such a common control model for power network monitoring field devices [17, 18]. For instance, DNP3 [19] was among the first protocols to have been designed to offer an open standard control system for power networks. DNP3 was based on the paradigm of exchanging generic control data while respecting the bandwidth constraints of the physical links. As such, DNP3 protocol could work with a minimum control overhead allowing it to run over communication channels with operating bit rates as low as 1200 bits per second by adjusting the size of the messages sent by control devices and reducing their sampling rates [16].

More recently, the International Electro technical Commission (IEC) proposed a new standard that targets interoperability and fast communication between Intelligent Electrical Devices (IEDs) called IEC61850 [20, 21]. The superiority of IEC61850 over other communication standards in substations is that the functions, the services, and the communication protocols are not mixed together but are defined separately. Moreover, the standard contains the data models of all possible functions in a substation, and standardizes the names of these functions and their data. Similarly, the standard specifies a set of generic abstract services which cover all the data transfers required within a substation and maps these abstract services and the standardized data onto real communication protocols which include Ethernet, TCP/IP and Manufacturing Message Specification (MMS) [22]. A study in [23] show that IEC61850 can use a common data model in substation automation even over low bandwidth networks, by adjusting the control messages sizes and the sampling rates of the IEDs. Hence, coordination traffic, and the field-device data traffic has relatively lower bandwidth requirements. However, recall that distributed control systems are projected for use in future power networks; hence they need to be extensible for future control functionalities. This implies that the communication solution that will be deployed to carry this traffic needs to be scalable both in cost and in performance.

2.4   Conclusion and Open Issues

The smart grid will be characterized by its ability to support heterogeneous energy supply with a mixture of traditional large power stations, and small renewable energy sources. The management of the power flow in this model will therefore become increasingly challenging and will require the modernization of the control and communication systems that govern the power grid. Active control aims to replace the existing passive control systems that govern today's power grid and providing better management of these energy flows. This control paradigm, however, is predicated on the dense deployment of sensors, and seamless coordination between autonomous controllers.

In particular, the implementation of active control for distribution and generation necessitates the introduction of changes in the communication infrastructure that supports control operations in the power grid. This communication infrastructure cannot currently support the control traffic generated by thousands of extra sensors and control devices that will be added to control the power grid. The study published in [24] shows that adding active control systems to manage small distributed generators increases the grid communication traffic overheads which, without proper management, may result in congestion and message loss which in turn will potentially threaten the performance of grid control operations.

Therefore, the existing communication infrastructure first needs to be urgently upgraded in terms of transmission capacity to support this extra control traffic. In addition, new communication links will need to be added in order to connect intelligent controllers and support the coordination of traffic. An upgrade of the communication infrastructure, however, will necessitate a full study of existing communication technologies in terms of performance, ease of deployment, and cost.

References

1. D. Bailey, and E. Wright, *Practical SCADA for industry*. 2003: Newnes.
2. Y. Sun, and N.H. El-Farra, *Integrating control and scheduling of distributed energy resources over networks*, in Advanced Control of Chemical Processes, vol. 7, n. 1: pp. 141–146, 2009.
3. P. Lopes, et al., Integrating Distributed Generation into Electric Power Systems: A Review of Drivers, Challenges and Opportunities, in Electric Power Systems Research, vol. 77, n. 9: pp. 1189–1203, 2007.

4. T. Niknam, A. Ranjbar, and A. Shirani. Impact of Distributed Generation on Volt/Var Control in Distribution Networks. in IEEE Power Tech Conference, Bologna, Italy, 2003.
5. T. Ackermann, and V. Knyazkin. Interaction Between Distributed Generation and the Distribution Network: Operation Aspects, in Asia Pacific, IEEE/PES Transmission and Distribution Conference and Exhibition, 2002.
6. R. Currie, et al., Active Power Flow Management to Facilitate Increased Connection of Renewable and Distributed Generation to Rural Distribution Networks, in the International Journal of Distributed Energy Resources, vol. 3, n. 3: pp. 177–189, 2007.
7. R. Lasseter, and P. Paigi. Microgrid: a Conceptual Solution, in IEEE 35th Annual Power Electronics Specialists Conference, 2004.
8. F. Katiraei, and M. Iravani, Power Management Strategies for a Microgrid With Multiple Distributed Generation Unit, in. IEEE Transactions on Power Systems, vol. 21, n. 4: pp. 1821–1831, 2006.
9. F. Katiraei, M. Iravani, and P. Lehn, Micro-grid Autonomous Operation During And Subsequent to Islanding Process, in IEEE Transactions on Power Delivery, vol. 20, n. 1, 2005.
10. P. Järventausta, et al., Smart Grid Power System Control in Distributed Generation Environment, Elsevier Annual Reviews in Control, vol. 34, n. 2: pp. 277–286, 2010.
11. Owonipa, A., M. Dolan, and E. Davidson, *The Need for an Agent Arbitration Approach for Coordinated Control in Active Power Networks*, in *46th International Universities' Power Engineering Conference* 2011: Soest â Germany.
12. M. Fila, et al., *Coordinated Voltage Control for Active Network Management of Distributed Generation*, in Power & Energy Society General Meeting, Brunel Univ, Uxbridge, UK, 2009.
13. F. Melo, et al., *Network Active Management for Load Balacing Based in Intelligent Multi Agent System*, In CIRED 2012 Workshop Integration of Renewables into the Distribution Grid.
14. T. Logenthiran, and D. Srinivasan, *Multi-Agent System For The Operation of An Integrated Microgrid*, in Journal of Renewable and Sustainable Energy, vol. 4, n. 1: pp. 1–20, 2012.
15. H. Helin, and M. Laukkanen, *Towrads Efficient and Reliable Agent Comminication in Wireless Environments*, in Cooperative Information Agents V, Springer, pp. 258–263, 2001.
16. G.R, Clarke, De. Reynders, and E. Wright, *Practical Modern SCADA Protocols: DNP3, 60870.5 and Related Systems*. Newnes, 2004.
17. C. Ken. "A DNP3 Protocol Primer." DNP User Group Calgary, Canada, 2005. http://www.dnp.org/AboutUs/DNP3 Primer Rev A.pdf.
18. IEC *IEC Energy Management System Application Program Interface (EMS-API)-Part 301: Common Information Model (CIM) Base*. Document IEC 61970–301, 2005.
19. Cent-A-Meter: http://www.centameter.co.nz. Accessed August 2014.
20. IEC, *IEC Communications Networks and Systems in Substations*, Document IEC 61850, 2005.
21. R. Mackiewicz, *Overview of IEC 61850 and Benefits*, in 2006 IEEE PES Power Systems Conference and Exposition.
22. C. Hoga, and G. Wong, *IEC 61850: Open Communication in Practice in Substations*, IEEE Power Systems Conference and Exposition, 2004.
23. T. Sidhu, and Y. Yin, *Modelling and Simulation for Performance Evaluation of IEC61850-Based Substation Communication Systems*, in IEEE Transactions on Power Delivery, vol. 22, n. 3: pp. 1482–1489, 2007.
24. F. Bouhafs, *Communication Technologies for Active Control in Distributed Generation Networks*, in International Journal of Distributed Energy Resources and Smart Grids, vol. 9, n. 2: pp. 183–209, 2013.

Chapter 3
The Smart Grid in the Last Mile

Control of the power grid, its equipment, and its energy supply operations is paramount to the safety of the infrastructure and its reliability. As we have seen, power network operators rely on SCADA systems to monitor the power grid and these systems were built over dedicated communication links and control equipment. The density and complexity of the distribution network that connects consumers to the power grid, and the cost of this dedicated communication and control infrastructure affects the level of investment on the modernization of the power grid. Indeed, current investment in power grid automation is focused on capital intensive equipment with transmission stations and generation receiving most of the attention. However, these factors resulted in a limited penetration of the SCADA systems beyond the core of the network.

This lack of interaction and communication between the energy operator and the consumers thereby denies the operator the opportunity to monitor the quality of the energy supplied, and the ability to dynamically react to power outages and supply disruptions. For instance, currently, power grid operators cannot detect power outages in real time and have to wait long periods of time, perhaps until it is reported, before realizing that some of their customers are without electricity. Moreover, the lack of a channel that allows real time communication between consumer and the energy supplier, denies the former the opportunity to obtain information regarding changes in electricity tariffs. In the future smart grid, where the energy market is predicted to be increasingly deregulated, information regarding changes in electricity price will be very important for the user so they can select the most cost-effective supplier.

Therefore, the modernization process at the edge of the smart grid will necessitate a radical change in the way utilities interact with their customers. Through the design of a communication and control infrastructure that will connect the grid operators to their customers, operators will be able to collect a wealth of information regarding the quality of their energy supply and the performance of the power grid. In return, consumers will get a more robust service and more information about what they are being charged. This chapter will explore existing technologies in the home and how the introduction of the smart grid will advance this.

© The Author(s) 2014
F. Bouhafs et al., *Communication Challenges and Solutions in the Smart Grid,*
SpringerBriefs in Computer Science, DOI 10.1007/978-1-4939-2184-3_3

3.1   Metering of Electricity

Electrical metering is the process that involves measuring energy usage at the point of consumption and the rate of this usage over fixed periods of time, often in watt-hours. By installing electricity meters within customers premises, utility companies are able to provide bills that show to the customers their exact energy usage, etc. The deployment of electricity meters in houses and buildings started with the spread of electrical energy at the end of nineteenth century and, although this process has been in use for more than a century, it remains largely based on the same technology. The smart grid will modernize the way utilities gather information about consumers' energy usage and how this is presented to consumers.

3.1.1   Traditional Metering

The most common type of electricity meters traditionally used by utility companies is the electromechanical meter that uses induction to measure electricity usage. In this model, a metallic disc is used to measure energy usage by counting the revolutions of this disc as it rotates at a speed proportional to the power passing through the meter [1], as shown in Fig. 3.1. Since the 1980s, meters available for common use have evolved from (1) electromechanical mechanisms driving mechanical, geared registers to (2) the same electromechanical devices driving electronic registers to (3) totally electronic (or solid state) designs. All three types remain in wide use, but the industry is now slowly moving towards solid state devices.

These electronic meters contain no moving mechanical parts—rotors, shafts, gears, or bearings. They are instead built around large-scale integrated circuits, other solid state components and digital logic, as shown in Fig. 3.2. Such meters are much more closely related to computers than to electromechanical meters. This eliminates the need for some of the mechanical complexity inherent in the geared mechanical registers.

The process of measuring energy usage in an electronic meter is very different than that of an electromechanical meter. Electronic circuitry samples the voltage and current waveforms during each electrical cycle and converts these samples to digital quantities.

However, although these solid-state electronic meters provide more accurate measurements, they still require periodic (manual) monitoring from energy utilities This means that agents must visit the customer premises periodically to retrieve meter's readings as they lack automatic reporting, which is costly for utilities as the number of meters is in the order of millions. In addition, the information displayed by these electronic meters does provide helpful feedback to the consumer, often just displaying a string of numbers that does not allow them to manage his energy usage more efficiently. For instance, a consumer would like to know how much it costs him to operate his washing machine, and how much energy this appliance is consuming in real time.

Fig. 3.1 Electromechanical meter

3.1.2 Smart Metering

Although modern solid-state meters represent a major improvement over traditional meters, utilities are still facing the challenge of measuring and monitoring the quality of energy supplied to consumers, especially as the number of their customers increases. As such, this process requires a number of labour-intensive operations, such as manual meter reading, field trips for service connects and disconnects, on-demand reads, power outage and restoration management, and other metering support functions, that could be handled more efficiently via automation, which is the driver for the new smart meters [2].

Fig. 3.2 Solid-state electronic meter

In addition to being solid state meter, smart meters also have built-in communication capabilities that allow it to send its measurements to the utilities directly.
Smart meters are one of the first smart grid remote communication technologies
and early installations record premises energy consumption at regular intervals and
use Automatic Meter Reading (AMR) [3] to send measurements periodically to the
utilities billing system. AMR systems can use a variety of communication technologies to connect with the billing system, such as: mobile communication (2G, 3G),
telephone line/ADSL, etc. and therefore represents a cheaper option to utilities as it
allows collecting information about consumers' energy usage as well as the status
of feeders and other equipment at minimum cost. This timely information coupled
with analysis can help both utility providers and customers' better control the use
and production of electric energy. The only limitation of AMR is that it provides one
direction communication: from the consumer's meter to the utility company.

3.1.3   Advanced Metering Infrastructure

Moving beyond these initial deployments, next generation smart meters will include the Advanced Metering Infrastructure (AMI) [4] and provide a two way communication channel between the consumers and utilities. Unlike AMR, AMI will
not only allow utilities to access smart meters' readings but will also provide scalable data collection, storage, and analysis mechanisms. Moreover, AMI systems
will allow utilities to send pricing signals to alert customers of peak pricing periods.
Such direct communication to the customers will encourage conservation during
peak periods and will enable utilities to implement more direct control of demand
side management.

Through near real-time price signals, AMI will also enable consumers to manage
their energy usage much more efficiently. These prices could be relayed directly to
appliances, or through a home energy management gateway as depicted in Fig. 3.3.
The appliances, in turn, can then process the information and adapt their performance based on consumers' performance to adjust power consumption accordingly.

Power Grid Operator

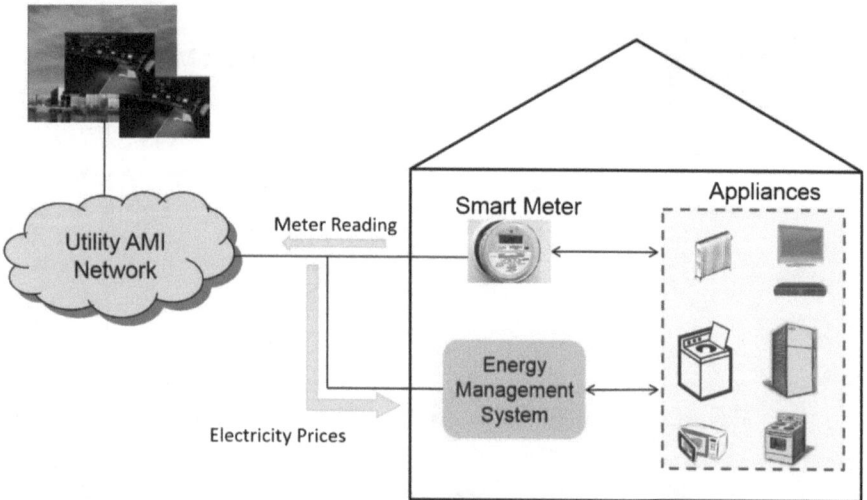

Fig. 3.3 AMI interaction with smart homes

AMI technology will therefore be a vital component of the smart grid as it will provide utilities with a wealth of new information that could be used to optimize business operations in addition to collecting monthly consumption data used for billing, provide load profile data, demand, time-of-use, voltage profile data, and power quality data.

AMI typically consists of a smart meter equipped with a communication interface that connects to the meter. Smart meters in households within the same neighborhood, equipped with communication interfaces, will then be connected to a central unit, called a data collector, forming a new Neighborhood Area Network (NAN) [5]. Each data collector will be connected to the AMI Wide Area Network (WAN) [6, 7], also called the Backhaul via this NAN, as illustrated in Fig. 3.4. A detailed description of these networks, their roles, and their characteristics will be presented in Sect. 3 of this chapter.

3.2   Demand Response System

Demand-Response (DR) [8] is another smart grid application that interacts directly with the consumer's electricity system, appliances, and other electrical devices within the premises. This application helps appliances to detect the status of the power grid, and if the current electricity supply to the appliances is during a peak period or not. This will allow appliances, or the home management system that control these appliances to adjust electricity usage such as postponing certain tasks according to the notifications sent by the DR system.

Power Grid Operator

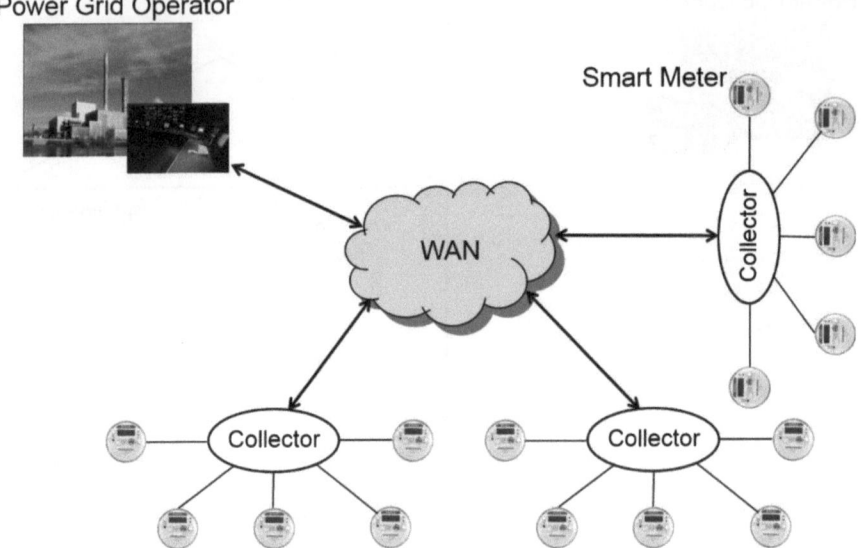

Fig. 3.4 AMI communication network

DR will be also used to manage residential energy generation. DR will allow the utilities to send notifications to these units to curtail, disconnect, or connected their generation to the grid according to the grid load and current energy demand. This will help to maximize the efficiency of local generation and avoid an overload if certain lines are reaching their capacity, hence protecting the electricity grid and increasing its efficiency and reliability.

3.3   Communication Architecture for AMI

Smart meters, DR, and the AMI are among the emerging technologies proposed to support the Smart Grid initiative. These technologies will allow utilities to implement customer-centered services and allow customers to interact with their suppliers more directly and dynamically. The implementation of these technologies requires the deployment of spur or last-mile communications typically from a backbone node right up to the customer premises. This requirement becomes more challenging once one considers that there are millions of homes and smart meters that need to be connected to the core communication network of the power grid.

As such, one scalable communication approach that could be adopted in the smart grid is to group Smart meters within the same neighborhood around an aggregator device, called a data collector, forming a Neighborhood Area Network (NAN)

[6]. Each data collector can then be connected to the AMI Wide Area Network (WAN), also called the Backhaul, via a collection points on the edge of the WAN providing connections and/or consolidation for metering data access as shown in Fig. 3.4. All data sent by smart meters and other control devices can then be relayed to the central server located at the utility control system, and which runs the management application.

3.3.1 Neighborhood Area Network

Typically, a NAN is a communication network that connects devices located in close proximity to each other although not necessarily connected to the same physical network, and not sharing the same broadcast domain. This type of network is useful for applications that need a scalable approach to connect thousands of devices densely located, as is the case with smart meters in the context of the smart grid. Although there is no standard that defines the structure and components of a NAN, it usually compromises a one of more base stations connected to a number of Home Area Networks (HAN), similar to Peer to Peer networks (P2P) [9].

AMI applications could therefore use a NAN to connect smart meters located in neighboring premises to local access points and these smart meters could be connected directly to the local access points, or in multi-hop using gateways, hence forming a mesh network. This type of mesh network is already being used in power grids to collect monitoring data from remote control devices forming what is called a Field Area Network (FAN) [10].

3.3.2 AMI Wide Area Network

Unlike NAN, Wide Area Networks (WANs) are more established networks that are operated by utilities and provide a link between substations and utility companies to monitor usage.

Although not part of the last mile communication network, the WAN provides an important functionality to last mile smart grid applications, as it connects the consumer electricity system to the smart grid control system. The first part of the WAN is called the Backhaul network and it connects the NANs to the utilities core network which represents the second part of the WAN.

The coverage of this network would be in the order of thousands of square miles while the data rates will be typical for large data networks, which are currently between 10–100 Mbps. In addition, the underlying technologies used may vary significantly based on the implementation and there is no defined technology or standard for this network. Therefore, the implementation of WAN for AMI could include a variety of candidates, such as Ethernet, GSM, etc. [11–13].

3.4   Communication Requirements and Solutions

Unlike SCADA systems which are limited to a few hundred collection points, AMI and other smart grid last mile applications will involve millions of customers spread across the distribution power network, mostly in residential areas. Although these applications provide a unique opportunity to gather more information on voltage, current, power, and outages, the challenge of deploying spur or last mile communications typically from the core network to all customer premises can be prohibitive. Utilities therefore face the challenge of designing and building communication networks that can support the connectivity requirements of smart grid applications, while at the same time finding the balance between the cost of such a process and the economic benefits of these applications.

3.4.1   Communication Requirements

The communication network for the last mile will therefore need to be offer scalability in both performance and cost. This requires a good understanding of the requirements of the Smart Grid applications, the deployment modes, and the data traffic characteristics.

In reality, residential smart grid applications, such as AMI, have relatively low reliability and connectivity requirements from the communication network compared to SCADA systems. For example, if a small number of customers are disconnected from the smart grid communication network for a relatively short period of time, the reliability and safety of the power grid operations would not be threatened. Moreover, these applications do not require high communication bandwidth as is the case in the backhaul. Therefore, cheaper, low-speed communication technologies with more marginal transmission bandwidth that may require multiple retransmissions to complete a message can be tolerated. These relaxed performance and reliability communication requirements in the last mile increase the number of technology candidates.

On the other hand, the communication network needed to connect customers to the backbone of their utilities will consist mainly of costumers' smart meters within the same neighborhood. As discussed above, smart meters in a neighborhood of around 1–10 miles^2 could easily form a mesh Neighborhood Area Network (NAN) by adding relay gateways, especially if using wireless communication. Such a network could cheaply and easily meet the smart grid requirements as most of the applications to be used in the last mile have low data rate 10–1000 Kbps.

3.4.2   Communication Technologies and Implementations

Despite intensive interest from the research community, there are currently still no clear standards for the implementation of the communication network for last mile smart grid applications, despite there being many possible wired and wireless communication candidates. On the wireless side, IEEE 802.11s, RF Mesh [12], Worldwide Interoperability for Microwave Access (WiMAX) and cellular standards, such as 3G, 4G, and LTE, are some of the stronger candidates [14]. On the wired side, Ethernet, Power Line Communications (PLC) or Data over Cable Service Interface Specification (DOCSIS) are possible options that could be used [11, 12, 15].

Recent contributions have now identified the components necessary for successful communication between the customer and the utility. In [16] the authors proposed the use of 802.15.4 radio using the ZigBee Pro networking stack along with the Smart Energy(SE) profile. In this type of implementation, smart meters will form a mesh network where data travels through a multi-hop path before reaching the AMI backhaul. This type of communication is suitable for urban areas where smart meters are densely deployed and the mesh network can provide a wide coverage but are perhaps less suitable in sparse rural environments.

Other implementations suggest the use of long rang wireless communications such as the Wi-MAX and 3G protocols [13]. In this type of implementation, smart meters will send their data directly to the AMI backhaul and negate the local aggregation as described above. However, in the case of 3G communication, it is also recommended that separate links are provisioned dedicated to the utilities operations only in order to increase security and avoid congestion or transmission delays. In this type of long range network, however, a costly investment might be required as utilities will need to build their own mobile radio base stations and equip smart meters with Wi-MAX radio antennas. As such, these solutions might only be used for remote rural areas where smart meters are sparsely deployed or located far away from the utilities backhaul network.

Similarly, there are currently no standards that define the topology and technologies to be used to implement the backhaul network that connects NANs to the utilities control system. However, there many possibilities to realise this network that are based heavily on current internetworking technologies including wired technologies such as: fibre channels, ADSL broadband and Power Line Communication (PLC), and wireless technologies such as: Satellite, Cellular, and Worldwide Interoperability for Microwave Access (WiMAX).

3.5 Conclusion and Open Issues

A lot of work is currently on-going to drive the development and deployment of last mile smart grid technologies, particularly because they promise to yield massive immediate benefits both to providers and consumers. Applications such as AMI and DR will help consumers to interact directly with their energy suppliers and involve them in the economics of demand and supply. However, this potential can only be fulfilled if communication channels are established between the consumer end systems and the utility backhaul network. Despite there being several candidates among the wired and among wireless communication technologies to do this, no definitive approach has emerged. Moreover, although these applications have very low reliability and delay requirements, the combination of low bandwidth offered, especially by wireless solutions, and the sheer number of devices connected to the utility backhaul network, creates a number of challenges.

Security is also a major concern in the smart grid as the communication between smart grid applications such as AMI and DR will certainly contain private information that could expose customer habits and behaviours if exposed [17]. For example certain activities, such as watching television, have detectable power consumption signatures that need to be protected in order to preserve the privacy of the consumers. However, the limited bandwidth offered by underlying communication technologies may still dictate the necessity to aggregate data generated by devices. This aggregation of control data, although helping to utilise scarce bandwidth efficiently, increase the chance of intercepting sensitive data and compromising the privacy of hundreds of consumers. The challenge therefore is to devise security solutions in which a trade-off is found between data privacy and transmission bandwidth.

Another challenge that needs to be addressed in last mile communication for the smart grid is reliability. Although data reporting traffic for these application does not have serious communication requirements, the underlying communication technology could affect the performance of the delivery as data is aggregated. For instance, routes in wireless technologies break frequently, perhaps as a result of fading effects and signal interference that makes the quality of wireless links highly unstable and time-variant. It is therefore necessary to utilise advanced routing protocols that could adapt to constant changes in the network topology and bandwidth. Also, to help implement these protocols it is necessary provide them with ability to uniquely identify elements in the network, and therefore the challenge is to devise an addressing scheme that covers all network components involved in the data delivery process, perhaps based on the new IPv6 protocol.

Finally, the communication infrastructures that support smart grid applications raise a number of further issues that need to be addressed. This infrastructure will be expansive and highly complex in nature and comprise many networking devices, such as gateways, relay, etc. potentially under the control of multiple providers. The challenge therefore is to devise effective and scalable management services to monitor the components of the infrastructure and provide fault detection, isolation, and recovery functionality to ensure its robustness.

References

1. H.M. Berlin, and F.C. Getz, *Principles of Electronic Instrumentation and Measurement,* Prentice Hall, College Div, 1988.
2. F. Benzi, et al., *Electricity smart meters interfacing the households.* IEEE Transactions on Industrial Electronics, vol. 58. n. 10: p. 4487–4494, 2011.
3. T. Khalifa, K. Naik, and A. Nayak, *A survey of Communication Protocols for Automatic Meter Reading Applications.* IEEE Communications Surveys & Tutorials, vol. 13, n. 2: p. 168–182, 2011.

4. D.G. Hart, Using AMI to Realize the Smart Grid, in IEE Power and Energy Society General Meeting-Conversion and Delivery of Electrical Energy in the 21st Century, 2008.

5. H. Gharavi, and B. Hu. Multigate Mesh Routing for Smart Grid Last Mile Communications, in IEEE Wireless Communications and Networking Conference (WCNC) 2011.

6. C. Bennett, and D. Highfill, Networking AMI Smart Meters, in IEEE Energy 2030 Conference, 2008.

7. D.M. Laverty, et al., Telecommunications for Smart Grid: Backhaul Solutions for The Distribution Network, in IEEE Power and Energy Society General Meeting, 2010.

8. S. Choi, et al., A Microgrid Energy Management System for Inducing Optimal Demand Response. in IEEE International Conference on Smart Grid Communications (SmartGridComm), 2011.

9. G. Fox, Peer-to-Peer Networks, in Computing in Science & Engineering, vol. 3, n. 3: p. 75–77, 2001.

10. Alcatel-Lucent. Field Area Network for Power Utilities: http://www.alcatel-lucent.com/power-utilities/field-area-network.

11. S. Galli, A. Scaglione, and Z. Wang, For the Grid and Through the Grid: The Role of Power Line Communications in the Smart Grid, in Proceedings of the IEEE, vol. 99, n. 6: p. 998–1027, 2011.

12. N.G. Myoung, Y. Kim, and S. Lee. The Design Of Communication Infrastructures for Smart DAS and AMI. in IEEE International Conference on Information and Communication Technology Convergence (ICTC), 2010.

13. P.P. Parikh, M.G. Kanabar, and T.S. Sidhu, Opportunities and Challenges of Wireless Communication Technologies for Smart Grid Applications, in 2010 IEEE Power and Energy Society General Meeting, 2010. IEEE.

14. M. Sugano, Integration of Different Smart Metering Systems Based on Wireless Communication. in IEEE International Conference on Distributed Computing in Sensor Systems (DCOSS), 2014.

15. S. Galli, A. Scaglione, and Z. Wang. Power Line Communications and The Smart Grid, in First IEEE International Conference on Smart Grid Communications (SmartGridComm), 2010.

16. S.W Luan, et al., Development of a Smart Power Meter for AMI Based on ZigBee Communication, in International Conference on Power Electronics and Drive Systems (PEDS), 2009.

17. P. McDaniel, and S. McLaughlin, Security and Privacy Challenges In The Smart Grid,.in IEEE Security and Privacy, vol. 7, n. 3: p. 75–77, 2009.

Chapter 4
Communication Solutions for Backhaul and Wide Area Networks

The novel technologies and services in the smart grid will be supported by more interaction between the control system and the electricity network than ever before. The introduction of distributed energy generation coupled with active control will enable more reliable energy supply and more effective and proactive management of the grid while smart meters and the AMI will allow utilities to better control the end-to-end quality of their electricity supply operation and interact better with customers. Central to this will be the modernization of control systems in the smart grid which will be based on the integration of advanced sensing technologies, distributed active control methods, and integrated communications into the current grid infrastructure. Control operations that in the future will rely on real-time knowledge of network conditions provided by the sensors to make coordinated real-time actuation instructions.

However, the current power grid lacks the communication capabilities to reach the level of penetration required by this new control system. Most of today's control system are limited to high and medium voltage parts of the power grid (transmission network)so the implementation of this system will necessitate the extension of the communication infrastructure to hundreds of thousands of control points located at lower voltage parts of the grid. Another issue that arises from the integration of advanced technologies and applications for achieving a smart grid infrastructure is the huge amount of data generated by different applications that will be used for further analysis, control and real-time pricing functions. As such, the scale and scope of the communication network are not the only factors that need to be taken into consideration, but also the bandwidth it offers, its reliability and functionality.

As we have seen, there are many technologies that can help realize the new smart grid communication infrastructure. These technologies can be classified as wired, and wireless and can all be used to build backhaul networks for communication between smart meters and utilities control systems, and for communication networks to support active control operations. This chapter will present these communication technologies, and their advantages and drawbacks with regards to the requirements of smart power grid and their control systems.

© The Author(s) 2014
F. Bouhafs et al., *Communication Challenges and Solutions in the Smart Grid*,
SpringerBriefs in Computer Science, DOI 10.1007/978-1-4939-2184-3_4

4.1   Wired Communications Media

Wired communication solutions were the first to be introduced in the area of power network control systems and have been the mainstay to date. Wired communication technologies range from low speed telephone lines to high-speed fiber optics and we will discuss the characteristics of each of these solutions, their advantages and their drawbacks.

4.1.1   Telephone Line

Telephone lines have long been used by electric utilities to meet their communication needs [1–3]. Leased (dedicated) lines are usually used for control operations that require connection with remote control substations as they allow permanent secure connection to remote site. However, the installation of these lines is usually mandated to telephone operators and requires the installation of new cables to remote locations. This task may be very costly in terms of the time and expense of deploying equipment at private properties, or may create long delays as lines may be installed over harsh terrain to the extent that, if many remote sites have to be connected to the SCADA system, this solution may become prohibitively expensive. Telephone (PSTN) dial-up lines on the other hand are cheaper as they require no installation beyond what a typical consumer requires. With these lines, the host can dial a particular number of remote sites to get readings or send commands. However, although the telephone network runs over a rich infrastructure of land lines it still does necessarily reach all areas particularly at remote sites where control substations may be located.

Telephone lines are generally built over twisted-pair metallic cables, as nearly all telephones are connected to the telephone exchange by a twisted pair. A twisted-pair metallic cable consist of two conductors that are twisted together to reduce the electromagnetic interferences from similar cables close by. Communication can be run over twisted-pair cables for several Km without the need for amplification, however; for longer distances repeaters are needed. The bandwidth offered by twisted pair depends on the thickness of the cable and the distance of the link. However, a twisted pair can offer a bandwidth of several megabits/sec for a link of a few kilometres.

4.1.2   Digital Subscriber Lines (DSL)

DSL allows the transmission of data over ordinary copper twisted-pair telephone lines at high frequencies, typically several MHz, providing broadband digital services. DSL collectively refers to the group of technologies that utilize unused capacity in the existing copper access network to deliver high-speed data communication

[2]. There are many variations of DSL, each aimed at particular markets, but they are all designed to accomplish the same basic goals. ADSL, or Asymmetric DSL, is a variation of DSL originally designed for the residential consumer market that can represent a cheap and reliable option for smart grid backhaul communication.

ADSL offers a high bandwidth that can go up to several Megabits per second. This bandwidth might be crucial for control traffic generated by control devices, sensors, and smart meters. ADSL can operate alongside existing telephone lines that are typically deployed to connect control sites to the control center and carry SCADA traffic, and requires little to no upgrading of the existing telephone lines used to connect power grid control sites to the control room. This means that the investment into the communication infrastructure will be very small. However, it is important to emphasize that when the control traffic is carried over a public network like the Internet, the security of the management platform might become a concern. The communication between the different system components can therefore be secured by using a Virtual Private Network (VPN).

4.1.3 Fiber Optics

Fiber optic communication was introduced in the 1960s as an alternative to the copper wire medium. Compared to other transmission cables, fiber optics offer higher bandwidth, smaller signal attenuation, and lower transmission delay. Very high data rates, as high as several gigabytes per second, can be transmitted over the fiber and fibers can be bundled together for even higher bandwidth. An optical fiber cable is a glass or plastic fiber used to transport light pulses from an end to end. Since fiber optics are based on light pulses transmission, its signal attenuation factor is very small. This characteristic makes fiber optics a good candidate for long distance communications and is now used extensively in Internetworking [2, 4, 5]. Furthermore, optic fiber is characterized by immunity to different interference sources such as Electromagnetic and Radio Frequency sources.

4.1.4 Power Line Carrier

Power Line Carrier (PLC) technology uses the high voltage power lines themselves as the communication medium. It offers the possibility of sending data simultaneously with electricity, therefore the only cost incurred by PLC is the cost of additional terminal equipment, as well as any intermediate repeaters. PLC uses a Line Matching Unit (LMU) to provide a connection to the high voltage transmission or distribution line and prevents the injected signal from spreading to other parts of the power network.

LMUs are usually implemented using capacitors. However, if the voltage transformer has a capacitive structure, it can also be utilized for line connection. A line

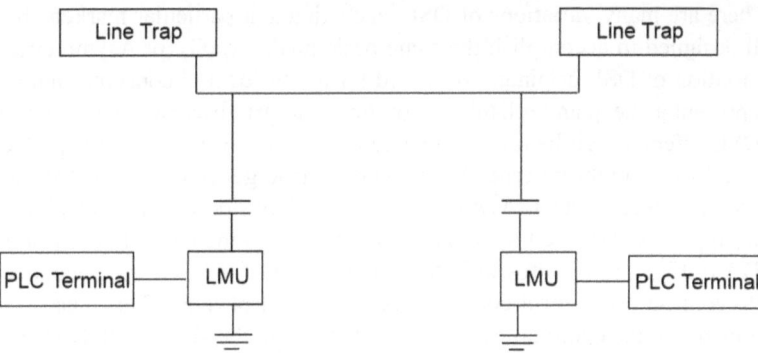

Fig. 4.1 Structure of a PLC System

trap is used where necessary to prevent the injected signal from spreading to other parts of the power network. The structure of a PLC system is shown in Fig. 4.1.

The standard IEC 60495 [6] defines the frequency band allocated for PLC as between 24 kHz and 500 kHz, and the width of a PLC communication channel as 4 kHz, with typically 64 kb/s per channel. By using a higher frequency, the PLC communication will require a smaller line trap inductance; hence the communication cost will be reduced. However, this higher frequency will induce higher attenuation limiting the communication span. In practice maximum transmission distances are a few hundred km. If longer distance communication is required, PLC terminals can be used to amplify the transmission signal as illustrated in Fig. 4.2.

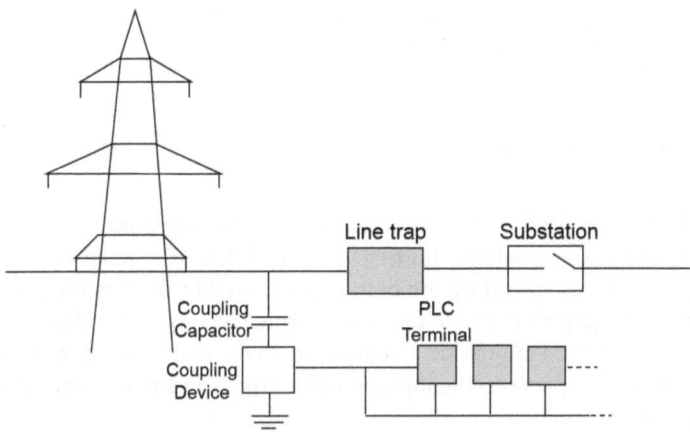

Fig. 4.2 Increasing the PLC Capacity

4.1.5 Discussion

Wired communication technologies present some attractive characteristics that make them a good transmission media for power network control applications, such as: immunity to electromagnetic interference, high bandwidth, and high reliability. However, the installation of cables typically comes at a high price, due mainly to the cost of digging and laying them. This represents a major challenge, especially if cables are deployed on a large scale or to connect field-devices deployed in remote locations to the control system. Moreover, the installation of cables might be impossible over harsh terrain such as mountains and hills. On the other hand, public communication networks such as telephone lines and DSL, require less work to install and are thus less costly, but they are usually not available in some rural places, include additional security considerations, and may not cover the entire power network.

PLC offers an affordable and practical solution for distributed control systems in power networks, as illustrated in Fig. 4.3, especially at secondary control substations which are located in remote places and are hard to reach with conventional wired media. However, the main technical drawback with using PLC over the distribution power network is the impedance mismatch resulting from the transition from an over-ground line to an underground line or the opposite. In case of more than one transition, the quality of the communication will be affected by reflections created

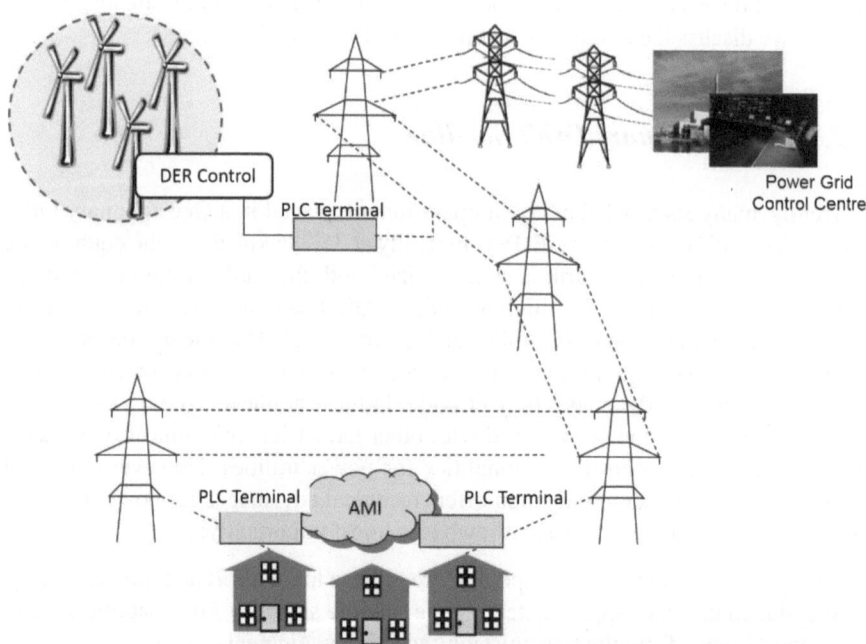

Fig. 4.3 PLC for Smart Grid Wide Area Network

by the mismatches resulting from the transitions. To overcome this problem, additional equipment needs to be installed at every transition point in order to reduce the effect of the mismatch, which will introduce significant cost. Therefore, before PLC is implemented, a full analysis of the distribution power network must be performed in order to determine the power line transitions. In addition, data quality in PLC is usually affected by the noise coming from power network equipment such as transformers, motors, etc.

4.2 Satellite Communication

Satellite communication technology has been used for many years in the domains of telecommunication and networking, and has been also adopted in power network control applications. This technology provides an extensive geographic coverage which makes it a good alternative in order to reach remote substations but they may be affected by atmospheric conditions. Currently, many existing services provide satellite communication varying from low to high data rates and for broadcast or two-way communications. Satellite communications can be considered as a microwave network with a big repeater in the sky, which provides greatly increased range. A typical communication satellite splits its 500 MHz bandwidth over a dozen *transponders*, and allocates 36 MHz to each transponder.

Each transponder can be used to encode a single 50 Mb/s data stream, 800 channels of 64 kb/s each, or any other combination of low and high rate streams. In this section we discuss the features of satellite-based communications.

4.2.1 Geostationary Orbit Satellite

Currently, many satellites that are in operation are placed in a Geostationary Orbit (GEO). The GEO satellite or GEOS is typically at 35,786 km above the equator and its revolution around the Earth is synchronized with the earth's rotation, hence its relative position is fixed. The high altitude of GEO satellites also allows them to cover approximately one third of the earth's surface [7]. The integration of GEOS communication has been proposed in the past for power networks control applications [8] due to the advantages they provide. In these applications, GEOS communication can work alone or in hybrid with other terrestrial communication systems in order to support control functionalities for power utilities. However, although GEO satellite based communication offers technical advantages for power network applications, it still presents some drawbacks, most importantly:

- Equipment installation: This operation requires a lot of effort and time to configure, due to the challenge of detecting the satellite signal, as GEO satellites are at a long distance from the transmission/reception equipment.
- Signal vulnerability to atmospheric and environmental factors.

- The large distance travelled by the signal from the sender to reach the receiver results in an end-to-end delay that varies between 250 ms and 300 ms. Added to this delay, is the 500 ms delay incurred by the TDMA based protocol used in satellite communication.

4.2.2   Low Earth Orbiting Satellite Communication

Low Earth Orbiting (LEO) satellites are typically deployed 200–3000 km above the Earth, which is significantly closer than with GEOS. This brings the round-trip delay down to 20–25 ms [9] which is comparable to the delay observed on some terrestrial communication technologies. Moreover, since LEO satellites are closer to the Earth surface, the necessary antenna size and transmission power level are much smaller and the installation operation requires less time and effort compared to GEOS. LEOS have already been successfully introduced to support a wide range of telecommunication services such as wireless Internet services, and could be used for power grid backhaul networks. LEO satellite based communication technology offers a set of intrinsic advantages such as: rapid connections for packet data, asynchronous dial-up data availability, reliable network services, and relatively reduced overall infrastructure support requirements if compared to GEOS [10]. In addition, LEO satellite based communication channels support packet oriented communication allowing protocols such as TCP/IP and PPP to operate, hence the possibility of carrying IEC61850 data traffic over these channels.

Currently, there are two main LEO communication service providers, namely Globalstar and Iridium. A study in [10] shows that Globalstar providers a better connectivity with a bandwidth that can reach up to 7.25 kbps, compared to 4.87 kbps offered by Iridium, at a lower Packet Loss Ratio (0% for Globalstar compared to 12% for Iridium). However, Although Globalstar exhibits better characteristics than Iridium, currently there are serious issues with the coverage and the communication integrity [11].

4.2.3   Discussion

Satellite communication offers good transmission bandwidth, high reliability, and wide coverage, which make it a good candidate for long distance communication between controllers, as in the case of an active control system. Moreover, unlike wired media, satellite communication can be deployed in remote locations that cannot be easily reached otherwise. The main drawback of this technology is the upfront cost of installing the communication terminals, assuming sufficient satellite capacity is available. LEO satellite communication represents a different price tradeoff to GEOS, as it requires cheaper equipment that is easier to install, however capacity is more limited so the data charges for regular communication can be prohibitive.

4.3   Radio Communication

Radio communication has also been used in power network control as an alternative to wired media to connect remote control entities that cannot be reached otherwise. Although radio communication cannot match the bandwidth offered by wired technologies, the reliability, performance and running costs of radio networks have matured considerably in recent years. Radio links therefore present a very tempting alternative technology for backhaul communication as it could help implement a number of smart grid application, as depicted in Fig. 4.4. The radio communication may be point-to-point or multipoint, operating typically either at UHF frequencies (between 300 KHz and 1 GHz) or microwave frequencies (between 1 GHz and 30 GHz) and this section will explore the options that currently exist.

4.3.1   Microwave Radio

Microwave radio refers to radio systems operating at frequencies above 1 GHz, offering high channel capacities and transmission data rates [1]. Microwave radio is widely used in long distance communication systems as an alternative to coaxial cable and fiber optics. Parabolic antennas can be mounted on masts and towers to send a beam to another antenna tens of kilometers away.

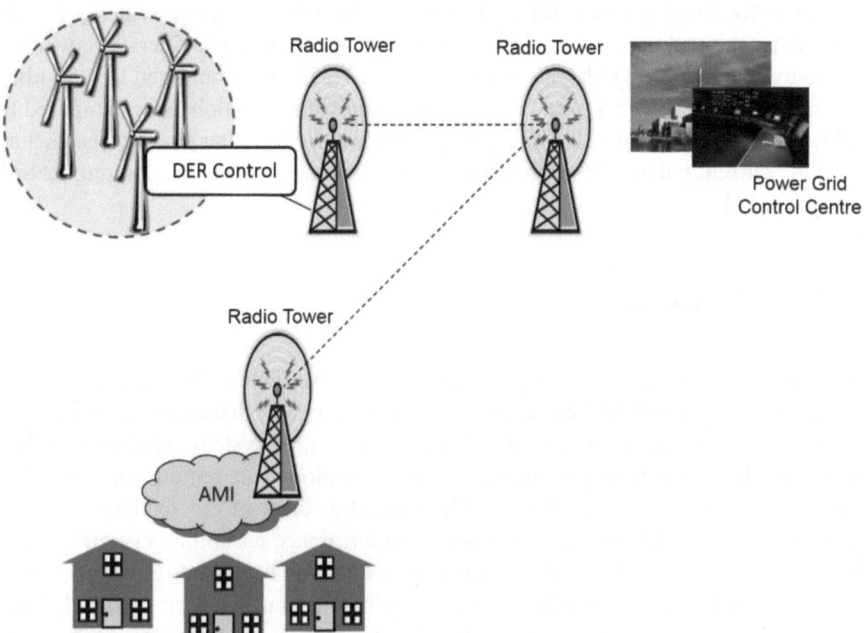

Fig. 4.4 Radio Technology for Smart Grid Wide Area Network

Microwave radio offers bandwidth which ranges from several Mb/s to 155 Mb/s. However, the capacity of transmission over a microwave radio is proportional to the radio frequency used so the higher the frequency, the bigger the transmission capacity, but the shorter the transmission distance. In addition, microwave radio requires line of sight between the two ends of the connection, hence, high masts are required. In the case of long distance communications, as is potentially the case between intelligent controllers, the installation of several high radio masts will be the major cost of microwave radio.

4.3.2   Ultra High Frequency Radio

Another alternative for wireless communication is the UHF-radio [1]. UHF radio designates a range of electromagnetic waves whose frequency is between 300 MHz and 1 GHz. Unlike Microwave radio, UHF does not require the line of sight, and therefore, the installation cost is decreased as high masts are not needed. UHF radio requires only a radio modem and an antenna and, depending on the size of the antenna size, it can easily span over 10–30 km offering a bandwidth that can reach 64Kb/s [12]. Moreover, the radio can also be used as a relay radio station, hence, expending the possible communication span. However, the number of hops may affect the capacity of the radio channel, i.e. more hops may decrease the effective transmission bandwidth.

4.3.3   Discussion

Radio communication alleviates control system designers from the burden of cable installation, and can help to connect remote entities to the control system. It is also cost effective as it reduces the cost of the communication infrastructure. Among all radio communication technologies, Microwave offers the highest bandwidth which makes it suitable for heavy data traffic. The main drawback of radio communication is the necessity to build high towers and install terminals that can extend the span of communication to reach remote control devices, hence, increasing the cost of the communication infrastructure.

4.4   Mobile Radio

During the last few decades, mobile radio networks have witnessed explosive progress with the development of the cellular phone network and mobile access to data. The GSM standard that has driven much of this development effort and, since the development of this standard, mobile systems and infrastructures have been rapidly built in virtually every country worldwide. Initially designed to carry voice only,

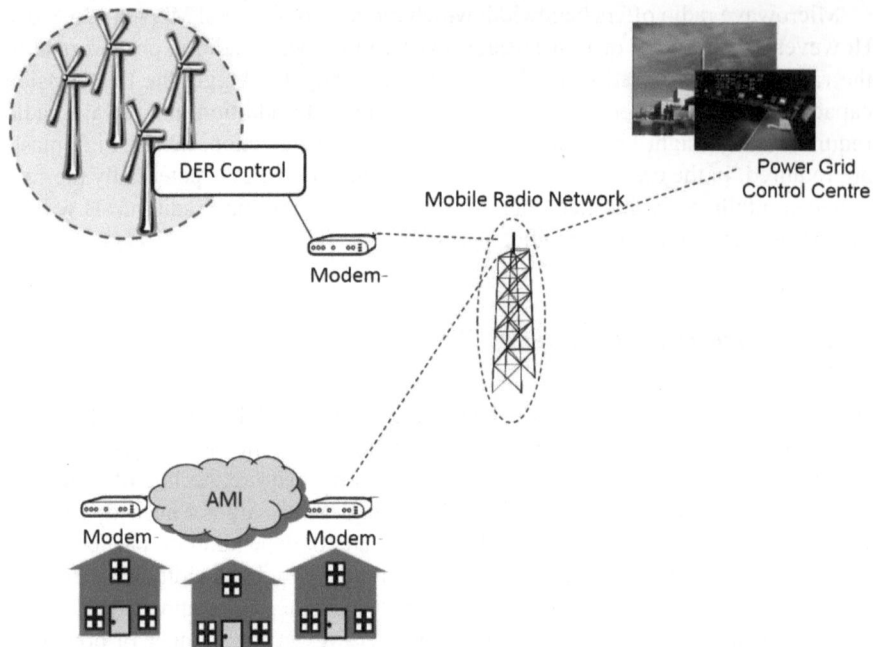

Fig. 4.5 Mobile Radio Technology for Smart Grid Wide Area Network

public mobile radio systems have evolved to support extensive data services to the extent that public mobile radio now has a performance that makes it a possible communication solution for remote control and monitoring applications [13, 14], as illustrated in Fig. 4.5.

4.4.1   Global System for Mobile communications (GSM)

GSM is an ETSI standard developed for the European public mobile telephony networks [15]. The GSM standard has known such success that operators throughout the world have adopted the standard. GSM offers both voice and short message service (SMS) data communication. The SMS data service supports message lengths of up to 160 characters, but four separate messages can be concatenated together. However, SMS messages are given low priority by the network and there are no guarantees of delivery. GSM also offers data communication services that are based on circuit switched technology with a maximum bandwidth of 9.6 kbps.

4.4.2 General Packet Radio Service

GPRS [16] can be considered as an extension of GSM mobile communication technology, specifically to support for data services. Although GPRS operates on the same radio bands as GSM (900 MHz and 1800 MHz) and shares the base stations with it, a different terminal is required to connect to a GPRS network. GPRS can be considered as an add-on data service to existing GSM technology with the difference that this service is always on. These '2G' services make GPRS much more suitable for wireless data communications than GSM with data rates up to 170 kbps.

4.4.3 Third and Fourth Generation of Mobile Phone Technology

Third generation and Fourth generation (3G/4G) [17] phone services takes the packet switched network a step further by providing a handset permanently connected to the network, hence removing the need for the connection initialization that is usually required in GPRS before transferring data. 3G/4G communication services are built around the data communication rather than voice as it is the case with GSM and GPRS systems. 3G/4G offer extensive data coverage, and since they operate at higher frequencies than GSM and GPRS (around 2 GHz), higher bandwidths (2 Mbps and 100 Mbps theoretical speeds) will be available with an improved quality of service.

Due to their extensive coverage, 3G/4G phones services could be used to connect very remote control substations to the main backhaul network as well as distributed generation controllers and houses smart meters. However, sometimes, establishing a connection to the cellular network can incur a significant time delay. In addition, 3G/4G services may not be an economically viable option if used for larger group of remote sites or regular data transfer [18].

4.4.4 Private Mobile Radio (PMR)

Unlike public mobile radio communication systems, PMR systems have been specifically designed for reliable professional use, which immediately makes them more suitable for power control systems. Similar to the progress made in public mobile radio networks, a standard for PMR has been developed by ETSI, called Terrestrial Trunk Radio (TETRA) [19]. Unlike GSM, TETRA targets only public safety and security applications as well as private companies control systems, such as utilities. Although being an open standard, TETRA network services are designed to be used in a closed environment which represents an advantage over public mobile radio GSM based networks. A TETRA system consists of one or more switches, one or more dispatch centres, base stations on fixed locations with good radio signal coverage, and user terminals (handheld and mobile radio stations) as depicted in Fig. 4.6.

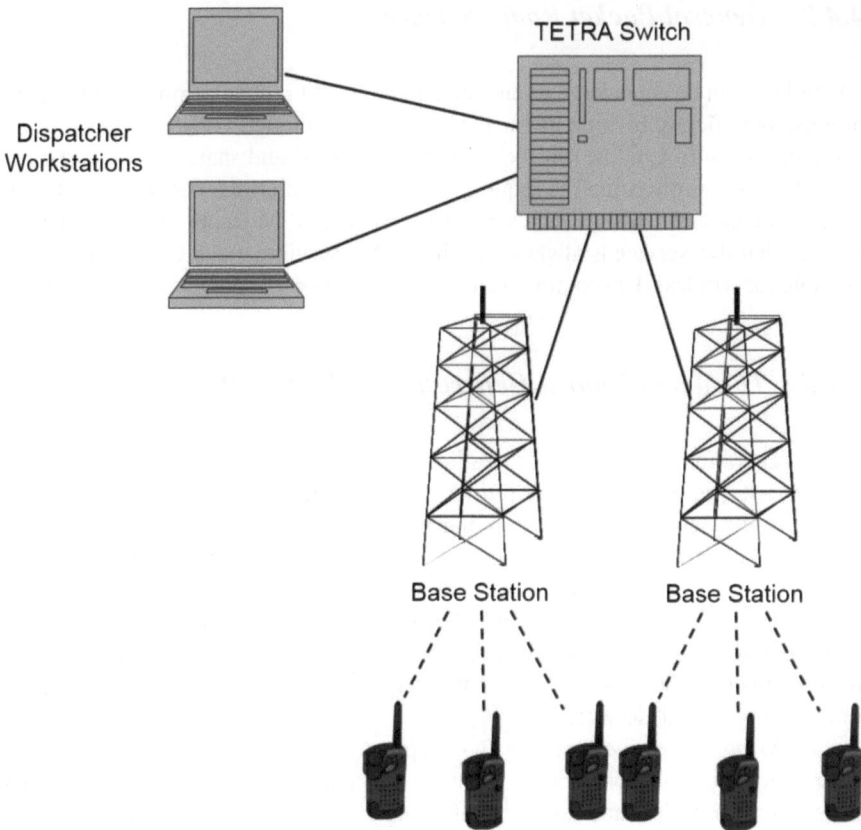

Fig. 4.6 Typical TETRA System Architecture

Each TETRA base station should use at least two carriers to achieve adequate reliability in case of breakdown in a single RF transmitter and related equipment. Thus, seven communication channels would be available for users (one is reserved for TETRA system use).

The system can be also be interconnected with other communication systems like wired and wireless phone systems and data communication systems via the IP protocol. TETRA operates in the 400 MHz frequency band, where the radio channel uses Time Division Multiple Access (TDMA) and the band is divided into 25 kHz channels each containing 4 different traffic channels.

This significantly increases the utilization of the radio channel (4 times more effective) compared to GSM. In addition, TETRA, like other modern digital PMR systems, supports virtual networks, enabling a number of independent users (organizations, enterprises) to share a common communication infrastructure (switches, base stations, etc.) whilst maintaining their integrity. In TETRA, data can be sent in packets (e.g. with the IP protocol) or in text messages by the Short Data Service (SDS).

4.4.5   Discussion

Mobile radio represents a cost effective communication solution for distributed control systems, as the cost of transmission/reception terminals is quite cheap. In addition, public mobile phone systems are now available almost everywhere, which makes them an excellent choice for the connection of control devices located in rural areas, and they can be deployed on a large scale. However, public mobile radio cannot be considered as a reliable communication media as they simply have not been designed for reliable operation in emergency situations and may be prone to congestion or failure. Thus, in critical circumstances when communication is needed most, they are likely to fail, partly due to overloading by public communication traffic, i.e. voice calls [20]. Moreover, the bandwidth available for data transfer varies according to the load of voice calls made over the network. In the busiest periods of the day, the available bandwidth may not exceed a few kbps.

Private mobile radio networks such as TETRA represent an alternative to public mobile radio as it is more robust, and has been specifically designed for emergency and control applications. Moreover, it offers secure communication channels, and allows also data encryption. However, although PMR offers higher bandwidth than public mobile radio, it is also more expensive than public mobile radio.

4.5   Conclusion and Open Issues

With the arrival of the smart grid, utilities face the challenge of modernizing the way they interact with customers and manage the energy supply process. Novel applications such as AMI and modern control systems will help utilities to obtain an accurate view of the health of the grid but introduce significant additional bandwidth and scalability requirements as the need to communications expands. However, existing communication infrastructure still reflects the nature of current control systems that govern the power grid, which are largely designed following the same methodologies as the early grids of the 1930s. Current communication infrastructure therefore cannot provide the extensive coverage and penetration required by new smart grid applications, they do not provide the vertical communication capabilities that reach sensors at low voltage grid, or AMI's NAN data collectors and nor do they support the horizontal communication needed to implement distributed control for heterogeneous energy supply.

The modernization of the power grid, therefore, will necessitate the modernization of the communication infrastructure. Utilities will need to extend this infrastructure significantly and introduce new communication technologies in order to satisfy the requirements of the smart grid applications in both transmission scope, and bandwidth. This operation, however, first requires an in-depth understanding of all communication technologies available, their installation cost, technical advantages, and their suitability for smart grid applications.

Table 4.1 Advantages and drawbacks of main communication trends

Technology	Advantages	Drawbacks
Telephone line	High bandwidth	Expensive deployment cost
Fibre optics	High bandwidth and low transmission delay	Expensive deployment cost
Power line communication	Extensive coverage Cheap deployment cost	Low bandwidth High signal attenuation Error prone
Microwave	High Bandwidth Easy to maintain Ideal for communication Over difficult terrain	Susceptibility to weather effects Requires clear line of sight Susceptibility to radio Interference
Ultra high frequency radio	Cheap deployment cost	Susceptibility to weather effects Susceptibility to radio Interference Low bandwidth
Satellite communication	Global coverage Easy installation	Long transmission delay Susceptibility to weather Effects High operating costs
Mobile radio	Low deployment cost Fast and easy installation	Low bandwidth

In this chapter we presented the main communication trends available in the marketplace and we highlighted the technical advantages and drawbacks of each communication medium, which are summarized in Table 4.1.

The study presented in this chapter shows that many communication technologies have the potential to satisfy the requirements of smart grid applications if deployed properly.

Through a combination of wired and wireless, public and private communication technologies, utilities will be able to establish a modern network that could connect all different control entities and smart grid actors, and carry the data traffic generated.

However, although most modern technologies are designed to carry huge amounts of data, there will be always certain segments of the network in which the communication channel will not be able to carry this data load, or are too costly to sufficiently provision, resulting in potential bottlenecks. Therefore, after the provisioning of the communication network, the challenge will be to devise mechanisms that can process the volumes of data generated by sensors, control devices, smart meters, appliances, etc. Techniques such as local data storage and aggregation need to be explored and implemented to reduce the load on the more bandwidth constrained segments of the communication network where possible.

References

1. D. M. Laverty, et al., *Telecommunications for Smart Grid: Backhaul Solutions For The Distribution Network*, in 2010 IEEE Power and Energy Society General Meeting.
2. R.L. Freeman, *Telecommunication System Engineering*, 4th edition, Wiley, 2004.
3. V.C Gungor, et al., *Smart Grid Technologies: Communication Technologies and Standards*, in, IEEE transactions on Industrial Informatics, vol 7, n. 4: pp. 529–539, 2011.
4. V. Gungor, and F. Lambert, *A Survey on Communication Networks for Electric System Automation*, in the International Journal of Computer and Telecommunications Networking, vol. 50, n. 7: pp. 877–897, 2006.
5. A. Leon-Garcia, and I. Widjaja, *Communication Networks: Fundamentals Concepts and Key Architectures*, 2nd edition, McGraw-Hill, 2004.
6. *IEC 60495, Recommended Values for Characteristics Input and Output Quantities of Single Sideband Power-Line Carrier Terminals.*
7. Y. Hu, and V. Li, *Satellite-Based Internet: A Tutorial*, in IEEE Communications Magazine, vol. 39, n. 3: pp. 154–162, 2001.
8. D. Bem, T. Wieckowski, and R. Zielinsky, *Broadband Satellite Systems*, in IEEE Communications Surveys and Tutorials, vol. 3, n. 1: pp. 2–15, 2000.
9. V. Madani, et al., *Satellite Based Communication Network for Large Scale Power System Applications*, in iREP Symposium-Bulk Power System Dynamics and Control-VII, Revitalizing Operational Reliability, 2007.
10. A. Vaccaro, and D. Villacci, *Performance Analysis of Low Earth Orbit Satellites for Power System Communication*, Electric Power Systems Research, vol. 73, n. 3: pp. 287–294, 2005.
11. United States Securities and Exchange Commission, FORM 8-K Current Report, Washington, D.C, 2007.
12. F.H Raab, et al., *HF, VHF, and UHF Systems and Technology*, IEEE Transactions on Microwave Theory and Techniques, vol. 50, n. 3: pp. 888–899, 2002.
13. A.J. Wilson, *The Use of Public Wireless Network Technologies for Electricity Network Telecontrol*, in Computing and Control Engineering Journal, vol. 16, n. 2: pp. 32–39, 2005.
14. E. Ozdemir, and M. Karacor, *Mobile Phone Based SCADA for Industrial Automation*, in ISA Transactions, vol. 45, n. 1: pp. 67–75, 2006.
15. M. Mouly, M.B. Pautet, and T. Foreword By-Haug, *The GSM System for Mobile Communications*, Telecom Publishing, 1992.
16. T. Halonen, J. Romero, and J. Melero, GSM, *GPRS and EDGE Performance: Evolution Towards 3G/UMTS*, Wiley, 2004.
17. *International Telecommunication Union*, available from: http://www.itu.int/home/imt.html.
18. P.P. Parikh, M.G. Kanabar, and T.S. Sidhu, *Opportunities and Challenges of Wireless Communication Technologies for Smart Grid Applications*. in IEEE Power and Energy Society General Meeting, 2010.
19. I. Ozimek, et al., *Using TETRA for Remote Control, Supervision and Electricity Metering in an Electric Power Distribution System*, in WSEAS Transactions on Communications, vol. 7, n. 4: pp. 289–299, 2008.
20. V. Sempere, J. Silvestre, and T. Albero, *Remote Access to Images and Control Information of a Supervision System Through GPRS*, in first IFAC Symposium on Telematics Applications in Automation and Robotics, 2004.

Chapter 5
Home Energy Management Systems

The smart grid initiative aims to increase the reliability and diversity of energy supply in order to reduce pollution and provide cheaper and cleaner energy alternatives. However, these objectives cannot be met if electricity consumption in homes and commercial buildings are not allied with these principles. Currently, appliances and electrical devices consume a significant amount of energy daily that could be greatly reduced or even eliminated if these devices are equipped with smart energy management technologies.

Energy management solutions within homes and commercial buildings could help consumers to adhere to these core objectives of the smart grid initiative which will also result in substantial savings in electricity expenditure. To realise these solutions it is necessary to first identify and make users aware of the appliances responsible for the majority of electricity usage, and understand the consumption pattern of these appliances by means of distributed sensing. However, the deployment of sensors should not be limited to the monitoring of appliances energy usage, but can also be used to monitor the environment and context in which these appliances are operating. Environmental and contextual monitoring will help increase appliances awareness and facilitate the implementation of efficient energy management policies, for example, turning off unattended devices.

As we have seen, communications technologies will play a major role in the realisation of home energy management as it will be used to implement distributed monitoring and control of appliances. It will also be used to establish two-way communications between the consumer to their electricity supplier, hence, extending the smart grid into homes and commercial buildings. Technologies such as smart meters, and advanced metering infrastructure (AMI) enables the provision of real-time pricing information and other services to consumers. In this context, this chapter will study the communication requirements of home energy management systems (HEMS), and the technologies that could be used to satisfy these requirements.

© The Author(s) 2014
F. Bouhafs et al., *Communication Challenges and Solutions in the Smart Grid*,
SpringerBriefs in Computer Science, DOI 10.1007/978-1-4939-2184-3_5

5.1 Energy Usage in Homes and Buildings

Today, residential and commercial buildings are equipped with a wide range of electrical devices and appliances that are collectively responsible for a significant daily energy usage. According to the United States Energy Information Administration, residential and commercial energy usage represents 75 % of the total US electricity consumption, with appliances and lightning accounting for 60 % of residential energy usage [1]. Further, the International Energy Agency [2] estimates that residential energy usage in OECD countries contributes to 30 % of the total electricity consumption, and contributes to 12 % of energy generation related carbon dioxide (CO_2) emissions. The agency also estimated that residential electricity usage during the last decade has grown by 12 % and it will continue to grow to reach 25 % by 2020. The electricity consumption in residential and commercial buildings in Europe represents 29 % of the total electricity consumption. The electricity consumed by residential and commercial buildings in Europe is the second highest after the industry sector [3].

Although there are many types of appliances and other electrical devices responsible for the energy consumptions in buildings and houses, the main energy consuming appliances can be categorised as follows:

- Heating and Cooling: This category include all heating, ventilation, and cooling (HVAC) systems used to improve ambient conditions in homes and offices.
- Lighting System: Along with HVAC, lighting systems are responsible for the bulk of electricity consumption in these buildings.
- Daily Appliances: This category is defined as all electrical appliances except space and water heating, lighting, and oven and main hob (cooking).

These three categories are responsible for the bulk of electricity consumption in residential and commercial buildings on a day-to-day basis and it is estimated that a 10–15 % reduction in residential electricity usage in the United States will result in energy savings of 200 billion kWh per year, equivalent to the output of 16 nuclear power plants [4]. Therefore, understanding how these appliances and services operate and how they can be utilised better will help to devise strategies to consume energy more efficiently and make significant gains in energy usage.

5.1.1 Heating and Cooling

HVAC systems are an integral part of many residential and commercial buildings. According to the US Energy Information Administration [1], HVAC systems account for nearly half of the total electricity consumption in homes and offices. In the UK, heating is responsible for the majority of electricity usage in homes, and although there have been on-going efforts to make houses more heating efficient through insolation solutions, electricity consumption for heating purposes has grown from 58 to 61.3 % and studies predict this portion will keep increasing in the future [5]. In Europe heating is also the largest consumer of electricity, with an average usage of 150 TWh per year [3]. However, cooling systems are not very popular in Europe, with an average consumption of 17 TWh mainly focused around the Mediterranean regions, such as Greece, Spain, and Southern France [3].

Although, houses are now built with heat efficiency in mind, with better insulation, heating is still the main factor in electricity consumption in residential and commercial buildings and so more active energy management strategies are clearly required. One way to reduce the energy usage of these appliances is to simWply switch them off totally when nobody is at home. However, more intelligent approaches would; have them switched on/off according to a demand-response notification during peak hours, adjust the temperature/humidity in the room dynamically based on the number of occupants, and only disconnect the HVAC system from the power board when the house is unoccupied in order to avoid any energy leakage.

5.1.2   Lighting System

Lighting systems are also a major consumer of electricity in residential and commercial buildings, although the electricity consumed by these systems are dwarfed by the amount of electricity consumed by HVAC systems. According to the US Energy Information Administration, 461KWh is being consumed by lighting systems in residential and commercial buildings the US, which represents 14% of all residential electricity consumption and 12% of total US electricity consumption. In Europe, lighting represents around 10% of the residential electricity consumption, being the third main consumer after electricity for heating and cold appliances. Electricity consumption of household lighting was estimated to be around 84 TWh in 2007.

As with HVAC, lighting systems are steadily being made more efficient over time with better lighting technologies replacing traditional incandescent bulbs. Household lamp technologies now include LED, halogen lamps, self-ballasted compact fluorescent lamps, and to some extent, also single and double capped fluorescent lamps without integrated ballast and high density discharge lamps. There have also been efforts, particularly in commercial buildings to introduce automated systems to dim or turn off lighting when no-one is present.

5.1.3   Daily Appliances

As defined above, this category includes all appliances that we use daily in homes and building with the exclusion of heating, lighting, and cooking equipment. This means appliances such as refrigerators, freezers, washing machines, computers, etc.

The last 40 years has witnessed a sharp increase in energy consumption by this category of appliances with this increase largely due to the increasing diversity and popularity of appliances and electrical gadgets such as computers, game consoles, chargers, etc. Households rely on these appliances for more and more of their essential daily activities as well as for entrainment. For example, households commonly use refrigerators and freezers to store food, microwaves, toasters and kettles to cook it, televisions, DVD and set-top box devices for entertainment, and laptops, tablets, and mobile phones for communication. Moreover, the capacity of these appliances is continually increasing, and they are becoming common place not just in houses but also in commercial buildings.

A survey published in [6] found that, on average, 50 % or more of the electricity used in the homes surveyed was used for appliances. The survey suggests that 16 % of household electricity powers cold appliances (fridges and freezers), 14 % is used for wet appliances (washing machines and dishwashers), 14 % for consumer electronics, and 6 % for information and communication technology. Most of the appliances in this category perform operations of low priority that can be postponed to off-peak hours, or can be totally disconnected from the power board when not used in order to avoid any energy leakage.

5.2   Smart Home and Home Energy Management

Moving toward encourage better energy usage among consumers is one of the main objectives of the Smart Grid initiative. As seen in the previous section, houses and commercial buildings are important players in the consumption of electricity. More specifically, appliances and diverse electrical devices consume a significant portion of daily electricity generated in order to fulfill certain conditions or deliver a specific functionality. This highlights the need to devise energy management solutions to improve the efficiency of electricity consumption. These figures also indicate that there is the potential to save a significant amount of energy, if appliances within residential and commercial buildings could be managed more efficiently. The challenge, however, is to devise and implement these smart energy management solutions without affecting the performance of appliances and negatively impacting people's standard of living or their productivity.

5.2.1   Existing Energy Management Solutions

Until recently, energy management solutions have been extremely rudimentary with the electrical meter being the only device that provides any kind of energy monitoring functionality to the consumer. However, this trend is changing as consumers are becoming more interested in solutions that could help manage their energy usage and reduce their electricity bill. Recent studies show that the number of home energy management users around the world is expected to reach 63 million by 2020 [7].

Increasing consumers awareness is the first step towards the success of home energy management, and there are now many solutions that provide consumers with feedback about their kilowatt-hour consumption and electricity costs in real time [8, 9]. Solutions such as Aztech [10], Cent-A-Meter [11], and EML 2020H [12] record the pattern of electricity usage within a house and shows the current rate of consumption in statistical, graphical, and visual data formats. Other companies such as Microsoft and Google are also introducing similar solutions into the market.

Utility companies are also trying to increase consumers' awareness by offering them online services that allow them to see their energy usage on the web alongside their electricity bills [13–15]. For instance, British Gas offers its consumers an online energy tracking service, where a consumer can enter their electricity meter's reading, and receive the cost of the current electricity consumption. Furthermore, the customer can set energy consumption goals and alerts in the context of reducing overall electricity usage.

However, increasing consumer's awareness should not be limited to the energy usage of appliances, but should also cover the environment and context in which appliances are operating. This will provide monitoring and control information at a greater granularity, allowing users to make better informed decisions and utilities to anticipate and manage demand with greater intelligence. In addition, if these appliances are also sufficiently smart such as they become environmentally and contextually responsive, further significant energy savings could be made. This is the driver for the Home Energy Management Systems.

5.2.2   Home Energy Management Operations and Components

A smart home energy management system should enable households to effectively centralize the management of energy services in a house, provide them with functions for internal information exchange, and help to manage contact with the outside world. It will also help households to optimize their energy consumption around their living style, rearranging the day-to-day schedule to minimize usage, and thereby secure a high quality of living condition whilst enabling people to reduce bills from a variety of energy consumption in a house. Smart energy management operations involve turning off the appliances when they are unused, adjusting the level of essential services in the indoor environment in response to sensor data such as temperature and brightness, or automatically suspending certain appliance tasks to off-peak hours where possible.

With HVAC appliances for example, the energy management system must ensure that these appliances stay on as long as it detects presence in the room. It will also manage the appliances to ensure that the temperature and humidity levels within a room are adjusted for the occupants with the support of thermostats. However, it will also actively turn HVAC appliances off when the room is unoccupied for a certain length of time. For lighting systems, the energy management system could again adjust the strength of lamination for occupants based on consumer preferences and light sensor data with the support of dimmers and temperature sensors. It must also ensure to switch off the lights when the room is unoccupied for any

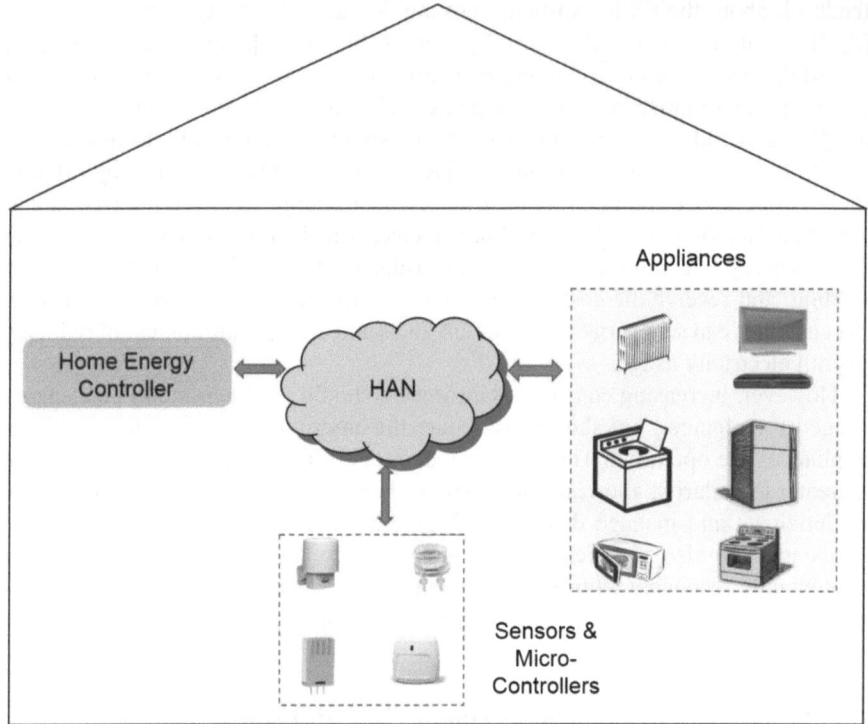

Fig. 5.1 Components of home energy management system

length of time. It might also utilise movement sensors to automatically activate the lighting when people enter the space. Finally, managing energy usage of appliances will mainly involve postponing or rearranging tasks according to energy cost at peak and off-peak hours.

An energy management system that could achieve these objectives will need to seamlessly integrate a range of technologies such as sensing, computing, and communication with a central Home Energy Controller (HEC) entity responsible for overall coordination, as shown in Fig. 5.1. This system will be in direct contact with appliances and other electrical devices via a Home Area Network (HAN), and will use sensors to continuously collect data and actuators to perform control and management operations throughout the building according to the energy management policy in place.

5.2.3 Smart Appliances

Appliances make a major and increasing contribution to the load on the power grid and as such are a very important part of any strategy for energy management. There are efforts to enable users' to change their consumption behaviour or to configure automated consumption thresholds, but this is not enough, so appliances must increasingly be proactively and automatically involved in the efficient management of electricity consumption.

Typically, a smart appliance combines embedded computing, sensing, and communication capabilities to enable intelligent decision-making and optimise its energy usage. The information obtained from the sensors increase the appliances awareness and allows them to react to new events that occur within their environment of operation. For instance, many households leave unattended appliances switched on, either deliberately or accidentally. A significant amount of energy could be saved if these appliances could detect the absence of a person in the room and react accordingly, much in the same way that a laptop, tablet, or smartphone already does. A Survey of 2,000 households shows the extent to which customers leave unattended appliances switched on, either deliberately or accidentally. Therefore, there is clearly a potential opportunity for Smart Appliances to avoid energy wastage, but there is a lack of reliable and cost effective occupancy detection sensors [16].

However, the contribution of each appliance depends on many factors, such as type of load, functionality provided, operation mode, environment and the context in which the appliance is deployed. Certain appliances could be considered smart if they operated in an energy saving mode and adapted to the variation of electricity prices by postponing and resuming their operation according to current electricity tariff. For instance, a washing machine could receive DR signals that the electricity tariff is high so it postpones its wash cycles until an off-peak period. A study conducted in [17] shows that this type of energy usage pattern result in a 25 % reduction in peak demand on the relevant section of the distribution network, and an 8 % reduction in average electricity consumption in participating households.

Unfortunately, many of today's appliances, water heating systems, and lighting fixtures, are not yet equipped with the required sensing, computing, and communication capabilities. These legacy appliances, therefore, cannot participate in the energy management operations without modification but this could be addressed in the short term by plugging legacy appliances into intelligent power outlets, called Smart Plugs [18]. Smart plugs are equipped with a sensor to measure the energy consumption in near real time, and communication capabilities hence allowing users to monitor energy usage and apply control remotely.

5.2.4   Environment Sensors and Actuators

For many appliances, increasing awareness also means relying on sensors to obtain information about the environment and context of operations in the building as this could yield significant energy savings. Sensing capabilities will enable the energy management system to measure energy consumption levels and gain report on environmental conditions in response. The role of sensing in an energy management system is to provide an environmental context in which appliances can validate

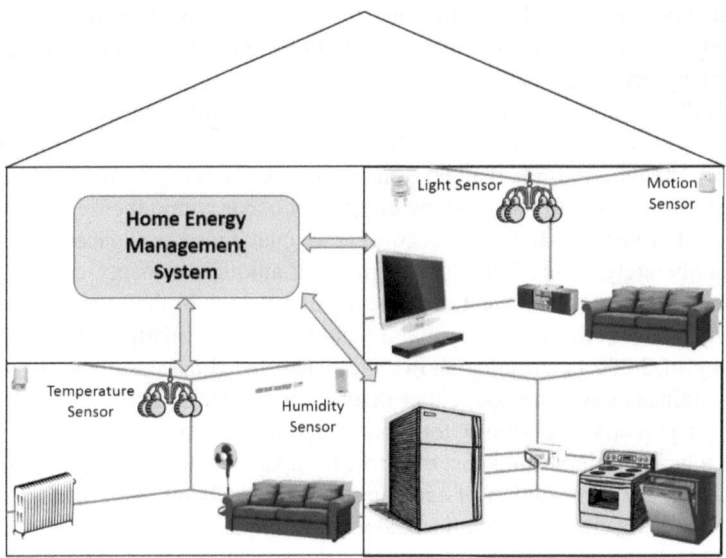

Fig. 5.2 Deploying sensors for home energy management

their mode of operation, for example adjusting the heating system according to a thermostat or turning off lights when there is no occupancy.

Although certain appliances such as HVAC systems possess these capabilities, their sensors provide a very limited view of the environment and the information provided cannot be considered reliable for energy management systems. For instance, an air conditioner could rely on its thermostat to determine if it reaches the desired temperature set by the user. While the data provided by the thermostat could help to determine the temperature in a relatively small space, it becomes less accurate in big spaces, e.g. at the far side of a room. Deploying sensor nodes within the house or commercial building premises will therefore increase appliances awareness and will help to make them operate more efficiently, and feed the home energy management system with accurate environmental information, as illustrated in Fig. 5.2.

A sensor node is a device that incorporates a number of sensors with some computation capabilities and a communication system. Sensors convert environmental stimulus into an electrical signal that can be measured stored, and reported back. There are a wide variety of sensor types available depending on the context, including: temperature, humidity, light, speed, acceleration, sound, magnetism. Moreover, the small size of sensor nodes and their ability to communicate wirelessly makes it possible to install them practically anywhere within the house.

Although sensor nodes are likely to be the main monitoring components of the energy management system, it is envisaged that these devices will work closely with actuators that operate on commands issued by the energy management system, in the same way smart plugs operate. The presence of actuators alongside sensors will help to increase the efficiency of the management system by allowing them to execute certain commands in response to stimuli that can range from simple on/off signals, to task postponing or turning an appliance into energy saving mode based on environmental inputs and other data.

In addition to sensors and actuators, tags could be used to extend the environmental context to indicate the presence of specified people or objects in a space, for example preventing children from turning on high consuming devices, or adjusting a fridge or freezer according to its contents. Tags are microchips that may be located in or carried by objects, animals or people to allow remote identification or tracking. Radio Frequency Identification technology (RFID) [19] comprises very small tags, composed of an IC and an antenna, and an RFID reader device. The reader transmits a radio frequency and reads the back scattered signal reflected and modulated by the tag to indicate its Id.

5.2.5   Home Energy Controller

Home Energy Controller (HEC) is a software application for managing energy-controllable smart appliances that will typically run on a central home server. It collects raw data from smart appliances, environmental sensors, and also smart meters to perform signal processing on the aggregate data and disaggregate energy usage. The HEC will provide detailed out to consumers and provide controls for them to automatically monitor and control home lighting, safety and security systems, and manage home energy usage. The HEC can also be used to schedule home appliances in order to reduce residential energy cost by avoiding peak energy pricing periods.

HEC performs device and consumption monitoring, allows configuration of energy thresholds, constraints and policies, and coordinates interconnection and meaningful communication between elements such as sensors, smart appliances, smart meter, generation systems and storage systems by executing the specified energy management policies. In some cases, a dialog could even take place between the smart appliance and the HEC to make appropriate decision according to the situation. For example, it might not be wise to interrupt a washing machine cycle if the cycle is about to end or turn off the lighting in a room if the person if heading back into the room.

The HEC could ultimately contribute to increasing utilities awareness about energy consumption within homes and buildings by also acting as a gateway between the utilities AMI and the home electricity infrastructure. The information reported by the HEC to the utilities could help in detecting the risk of dangerously high load or provide policy makers with environmental information along-side consumption information, enabling them to better direct energy saving resources. Since the exact composition of these devices has yet to be defined, in practice the HEC may be a dedicated device or a collection of functions across appliance automation displays, laptops, Programmable Communicating Thermostats (PCT) [20], etc.

5.3   Communication for Home Energy Management

Information sharing is paramount to the development of home energy management because, as we have seen, all the components in the system will need to communicate frequently to exchange data. Whether the HEC is implemented into a central device or distributed across different smart appliances, sensors and appliances will need to provide control information in order to determine and apply the appropriate energy management policy.

However, designing the communication network that will support this management system will need to take into consideration many factors. First, there are a range of control operations that must be supported, with each operation involving a different class of components and requiring a different communication performance. For instance, environmental monitoring within a room may involve infrequent but continuous communication between temperature sensors and the HEC only, whereas switching an appliance on or off involves the HEC sending a one-off command to the appliance itself. As such, the nature and operation mode of the components in the energy management system will dictate the nature of the communication technology that could be used. For instance, certain appliances such as a washing machine or dishwasher are installed at a specific location within the home and are not expected to be relocated and could therefore be connected easily via a wired solution such as an Ethernet cable. Whereas other components such as sensors may be located in hard to reach locations that allows them to provide accurate observations, and thus wireless communication is more suitable to connect these devices to the home energy management system. In this section, we will review in detail the characteristics of the communication networks on which a home energy management system could be implemented, identify the main requirements of these networks, and present a range of technologies that could help to implement them.

5.3.1   Communication Architectures for Home Energy Management

The communication networks involved in the energy management system could be expected to form a hierarchical architecture depending on the nature of the control operations and premise managed by the system.

At home level, the home energy management system will be expected to monitor and intelligently adjust energy usage by interfacing with these sensors, smart meters, appliances, smart plugs, etc. This process necessitates the integration a range of communication technologies forming a Home Area Network (HAN) [22]. This HAN therefore represents the platform for communication and internetworking of appliances, home energy controllers, and smart meters and can be built on top of existing data networks already in place. However, unlike other HANs that are used for home entrainment and multimedia application, this network will dedicated to energy management and will therefore only be used to carry control messages and commands between loads, smart appliances, etc.

Beyond this, it is ultimately expected that every building connected to the smart power grid would aggregate these HANs together to maintain its own Building Area Network (BAN) [23]. For example, a BAN might consist of a number of apartments in a block of flats or a collection of separate offices within a single building, each with their own HANs. The BAN smart meter would then act as the main gateway (GW) as it is typically set up at the building's power feeder. The BAN GW can then be used to monitor the power need and usage for the whole of the corresponding building. In order to facilitate BAN-HAN communication, Wi-Fi or ZigBee may be used to cover the area [27]. The upstream NAN could then connect BAN smart meters to local access points (the data collectors described in chapter 3) for AMI applications. The version of this network which is deployed to collect data from power lines, mobile workforce, towers, etc. for power grid monitoring is referred to as Field Area Network (FAN).

5.3.2   Communication Requirements

The design and realization of a home energy management system will therefore be based on the integration of many technologies such as: sensors, actuators/micro-controllers, wireless communication, etc. and although most of these technologies are already used in homes/offices, they have never been integrated on the scale that is required by home energy management systems. The main challenges in implementing the communication network to support energy management operations will be to connect all objects and entities to each other and to the external smart grid simultaneously. This communication network should offer the consumer global and granular monitoring capabilities of energy usage within their homes while also enabling utilities remote access to centralise the control of monitoring and metering.

However, as we have seen, given the diversity of the entities and nature of the control operations involved in the control process, the communication for energy management system is unlikely to rely on a single communication medium. In practice, the specific requirements will vary according to which elements are being connected. For example, connecting lighting to occupancy sensing in a home will have different requirements from connecting an appliance to a property meter in a large commercial building, or connecting a smart meter with a data collector or substation. Thus, we must also consider the properties of the elements themselves, such as functional lifetime, cost, form factor, and installation and configuration effort when selecting a communication medium. For example, most appliances are connected to the power infrastructure directly, but elements such as sensing may not and so must be able to manage power efficiently.

Given this diversity across the power network from source to appliance and given the collaborative aims of the smart grid, interoperability is an important, if not essential, quality for these technologies. It should therefore be considered whether the available communication technologies have the potential to support a common transport layer, such as that provide by the TCP/IP suite, and what affinity they might foster amongst mixed communication media and whether inter-working can be achieved with emerging power grid elements such as smart meters.

Finally we have a number of non-functional qualities, such as; the technology's potential for widespread (international) adoption and whether its design supports this ubiquitous deployment; the status of its standard, whether it is open and free and how widely it is certified; and the technology's potential for benefit to a wide range of stakeholders, including consumers, utilities and appliance manufacturers.

5.3.3   Communication Technologies and Solutions

We expect that a number of networking technologies will be involved to form the core of the communication network that will support a range of control operations in the HEMS. These networking technologies vary from high-speed core links to wired and wireless low power and rate limited links, such as: Wi-Fi, Power-Line Carrier, IEEE 802.15.4 and many others.

Historically, Power Line Communication (PLC) [24] has been a popular communication medium for appliance control due to its pre-existence in virtually every part of a house and a number of open and proprietary standards exist for PLC based data communications. X10 [25] is perhaps the best-known home automation standard due to its availability and simplicity. X10 enables plug-and-play operation with any home appliance and does not require special knowledge to configure and operate a home network. However as a carrier for energy consumption data or complex appliance control, X10 is limited by its data-rate which is typically around 20bps.

On the other hand, Ethernet, a less popular communication for appliance control, offers more reliable transmissions and higher data rates that vary between 1Mbs and 10 Gbs (typically now 100Mbs or 1Gbs). Ethernet runs over coaxial, twisted pair or optic fibre, and is now considered a very mature technology. However, due to the cost and inconvenience that might be caused by deploying additional Ethernet cables, the utilization of this technology is generally limited to a certain number of appliances such as game consoles, digital TV sets, etc.

The end of the last century witnessed the emergence of practical wireless data communication technologies such as IEEE 802.11 and Bluetooth that made it possible to connect devices together without the need for major installations into homes and building and at reduced costs. More recently, IEEE 802.15.4 [26], a new wireless communication standard has been proposed for low-power communication that characterizes many appliances and control devices. For that purpose, the ZigBee Alliance, an industry working group, is developing standardized application software on top of the IEEE 802.15.4, namely Zigbee RF4CE [27]. Zigbee RF4CE is a home-area network technology built on top of IEEE 802.15.4 that has been designed specifically to replace the proliferation of individual remote controls. This technology was created to satisfy the market's need for a cost-effective, standards-based wireless network that supports low data rates, low power consumption, security, and reliability.

5.4 Conclusion and Open Issues

Energy management systems for homes and commercial buildings represent a major component of the smart grid initiative. Such systems could help to reduce energy consumption significantly by maximising the efficient usage of appliances and resulting in reduced electricity bills. Current effort, however, is given to managing standby power and monitoring of overall energy usage which is not sufficient to identify if an appliance is consuming energy inappropriately.

Appliances therefore need to be equipped with capabilities that enable them to detect and manage their operation when it not contextually appropriate. Appliances will also need a uniform and ubiquitous support mechanism for obtaining energy management data which is self-configuring, ubiquitous, secure and fosters mixed participation. Key to the realisation of such a vision is the deployment of appropriate sensing and computation devices to implement a common energy consumption context, and designing a uniform appliance communication and actuation interface. This will enforce a data centric communication approach, as appliances are only interested in obtained control data from different devices (appliances, sensors, etc.) and not which device sent the data.

Current networking approaches such as TCP/IP, however, are mostly address centric where it is the responsibility of client nodes to discover the identities of nodes they need data from and set up connections between them. Although there have been efforts to make address centric approaches more light weight, communication still results in a significant overhead not suited to low resource devices such as sensors, and the needs of address-centric and data-centric approaches remain fundamentally mismatched. On the other hand, data centric communication has been used in the past in the area of wireless sensor networks that are mainly composed of symmetric, resource limited nodes, often with high mobility. The challenge here is to devise novel data centric communication approaches that address the heterogeneous nature of the home energy management network. It is also important to emphasize the importance of data naming in data centric communication as it focuses on a named data abstraction rather than a location abstraction. Generally, in a data centric communication network, each node autonomously establishes data forwarding rules based on the data's name rather than its destination, through a combination of interest, advert and reinforcement messages. It is therefore necessary to devise a data naming scheme that could help realise the data centric communication solution for the home energy management system. Finally the design of the uniform appliance communication and actuation interface will need to take into consideration the concerns of the particular stakeholders involved. There is already an understanding and acceptance within the appliance industry that we are progressing towards smart appliances, and research shows that there is already some willingness amongst consumers to accept smart operation in the home [28]. However, for appliance manufacturers it is essential that they protect the huge design and development investments, their market, and their consumers which may stifle rapid progress. To move the industry from concept acceptance to realization it will therefore be essential that their concerns are addressed. The primary concerns of appliance manufacturers are cost, added benefit to consumers, simplicity, stability and privacy. As the technologies and standards mature, we expect to see media integration, common transport, and support for a ubiquitous infrastructure, emerge as catalysts for the wide adoption and support for the energy management application domain.

References

1. US Residential Energy Consumption Survey (RECS), U.S. Department of Energy 2009, available from: http://www.eia.gov/consumption/residential.
2. *International Energy Agency, available from: http://www.iea.org/.*
3. P. Bertoldi, B. Hirl, and N. Labanca, *Energy Efficiency Status Report on Electricity Consumption and Efficiency Trends in the European Union. Status Report* 2012. European Commission, Joint Research Centre, Institute for Energy, Luxemburg (2012), available from: http://iet.jrc.ec.europa.eu/energyefficiency/sites/energyefficiency/files/energy-efficiency-status-report-2012.pdf.
4. *J. Froehlich, et al., Disaggregated End-Use Energy Sensing for the Smart Grid, in IEEE Pervasive Computing, vol. 10, n. 1: pp. 28–39, 2011.*
5. *J. Palmer, et al., , United Kingdom Housing Energy Fact File, Department of Energy and Climate Change, London, UK, p. 145, 2013.*
6. *P. Owen, Powering the Nation: Household Electricity Using Habits Revealed, Energy Saving Trust, London, 2012.*
7. *Pike Research, available from: http://www.navigantresearch.com/.*
8. *B. Parks, Home Energy Dashboards, in Make: Technology on your own time, vol. 18, pp. 84–51, 2009.*
9. *K. Roth, and J. Brodrick, Emerging Technologies: Home Energy Displays, in ASHRAE Journal, vol. 50, n. 7: pp. 136–138, 2008.*
10. *Aztech: http://www.generalpacifi*c.com/services/metering/aztech-in-home-display.
11. *Cent-A-Meter*: http://www.centameter.co.nz.
12. *EML-2020H*: http://www.powermeterstore.com/p4724/eml_2020h.php?thickbox†= images&currency = CAD.
13. *British Gas Energy Smart*: www.britishgas.co.uk/products-and-services/energy/our-taris/energysmart/how-energysmart-works.html.
14. *Google PowerMeter*: http://www.google.com/powermeter/about/about.html.
15. *nPower Energy Smart*: http://www.npower.com/campaigns/smartpower/index.htm.
16. *Delivering the Benefits of Smart Appliances,* Department for Environment, Food and Rural Affairs, EA Technology, London, UK, 2011, available from: http://randd.defra.gov.uk/Default.aspx?Menu=Menu&Module=More&Location=None&Completed=0&ProjectID=17743.
17. A. David, *Ethos Project, Esprit European Funding Programme, Overall Project Report*, Cardiff, 1998.
18. *Smart Plug*: http://www.alertme.com/products/smartplug-1622.html.
19. C.M Roberts, *Radio Frequency Identification (RFID)*, in Computers & Security Journal, vol. 25, n. 1: pp. 18–26, 2006.
20. T. Peffer, et al., *How People Use Thermostats in Homes: A review*, in Building and Environment Journal, vol. 46, n. 12: pp. 2529–2541, 2011.
21. ZigBee Alliance, *ZigBee Specification* 2008.
22. D. Niyato, L. Xiao, and P. Wang, *Machine-to-Machine Communications for Home Energy Management System in Smart Grid, in* IEEE Communications Magazine, vol. 49, n. 4: pp. 53–59, 2011.

23. Z.M. Fadlullah, et al., *Toward Intelligent Machine-To-Machine Communications in Smart Grid*, IEEE Communications Magazine, vol. 49, n. 4: pp. 60–65, 2011.
24. M.S. Yousuf, S.Z. Rizvi, and M. El-Shafei, *Power Line Communications: An Overview-Part II,* in IEEE International Conference on Information and Communication Technologies: From Theory to Applications, 2008.
25. D. Rye, *The X-10 POWERHOUSE Power Line Interface Model# PL513 and two-way power line interface model# TW523*, Technical Note, Revision 2, 2013.
26. E. Karapistoli, et al., *An Overview of the IEEE 802.15.4a standard*, in IEEE Communications Magazine, vol. 48, n. 1: pp. 47–53, 2010.
27. ZigBee Alliance, *ZigBee Specification* 2008, available from: http://home.deib.polimi.it/cesana/teaching/IoT/papers/ZigBee/ZigBeeSpec.pdf.
28. W. Mert, J. Suschek-Berger, and W. Tritthart, *Consumer Acceptance Of Smart Appliances,* Smart Domestic Appliances in Sustainable Energy Systems (Smart-A), 2008, available at: http://www.smart-a.org/.

Chapter 6
Communication Technologies for Smart Energy Management Systems

The implementation of smart energy management systems will rely on the integration of several technologies, including: sensing and actuating devices, smart appliances, some kind of controller, and the communication network. In the future, it is envisaged that a variety of sensing devices will be deployed within buildings in order to monitor the occupants and environment in which appliances are operating. Smart appliances will use the data provided by these sensing devices to adjust their operation mode, and hence manage their energy usage according to changes in the environment they are operating in. In certain situations, appliances will not be able to perform the necessary changes in isolation, and will need the intervention of actuators.

As described in the previous chapter, it is clear that the implementation of smart energy management systems will therefore be predicated on the availability of home communication networks that allow devices to share monitoring information and exchanging control commands. Given the diversity of available communication technologies, this chapter will identify the communication solutions and standards that could be used in the context of smart energy management systems, describing their characteristics, assessing their qualities, and identifying their limitations.

6.1 Power Line Communication (PLC)

PLC technology uses the existing power distribution infrastructure for communication between nodes which gives it a cost advantage over other wired technologies. PLC [1, 2] can use sockets for both power and communication and could provide a pre-existing line back to the Utility, making it a good candidate for ubiquitous appliance communication in buildings.

Though there can be performance issues over long distances and over complex legacy power networks, a number of open and proprietary standards exist for this medium [3]. However, while there is no doubt that new technology is beginning to manage power line problems better, this is still a legacy, unstable, and somewhat

© The Author(s) 2014
F. Bouhafs et al., *Communication Challenges and Solutions in the Smart Grid*,
SpringerBriefs in Computer Science, DOI 10.1007/978-1-4939-2184-3_6

hostilemedium often requiring the intervention of compensating devices. Moreover, since much of the technology here is proprietary and existing standards are limited, devices are location-restricted and the network is difficult to extend making this an unlikely medium for realizing context awareness.

6.1.1 X10

X10 [4] was among the first communication standards for home automation using PLC as its primary medium. However, the X10 protocol limits the transmission rate to 20 b/s, making it unsuitable for regular transmission of environmental sensing reports, and so it can only be realistically used for turning devices on or off or sending very simple instructions.

The X-10 protocol also provides limited security in the form of secure device addressing. This is manually configured on the device only during the installation state and is unlikely to be suitable for billing and other sensitive data. Moreover X-10 is not design to work well with common transport protocols such as TCP/IP. When considered with the above, these qualities make X-10 an unlikely candidate for wide scale adoption and it is unlikely to be seen as supporting the requirements of all energy management stakeholders.

Additionally a range of other users devices attached to the PL network can contribute to issues with attenuation, impedance and noise and the protocol itself presents issues of speed and collision. As such, X-10 networks commonly require phase couplers, blocking filters and signal boosters for larger power line networks, which further contribute to it failing to meet the smart gird goals of self-configuration and one-off easy deployment. Given its maturity, X-10 is also not significantly low cost and device form factor remains medium to large.

6.1.2 Ineston

The INSTEON [5] protocol has been proposed by home automation company Smart-home Technology to work on PLC. In this case, the network employs a non-routing topology where all nodes receive and repeat messages. In order to take advantage of common transport protocols, such as TCP/IP, Insteon devices must be fitted with purpose built serial interfaces such as USB, RS232 or Ethernet and connected with other digital devices which support bridging to other 'Insteon incompatible' networks such as a Local Area Network (LAN) or the Internet. Insteon achieves a sustained raw data 2,880 bps operating a dual medium (PLC and radio-frequency) peer-to-peer network.

Recently North America's biggest distributor for electricity utilities, 'HD Supply Utilities', will be offering packages of smart grid solutions to utilities, using integrated smart grid configurations of SmartLabs Inc's certified Insteon devices.

Offerings are expected to include demand response, home devices, AMI, management of meter data, and security.

However, Insteon's proprietary nature makes wide scale adoption unlikely. Through paid membership of the Insteon Alliance, one may have the opportunity to influence Insteon design and development, but its specification is not approved by European or International standards bodies and is copyrighted by and only available from Smart-home Technology. Dependency on such proprietary technologies may constrain ubiquitous deployment.

6.1.3   Lonworks

Lonworks [6] is a protocol developed by the Echelon Corporation [7] to operate on PLC, although it can also work on twisted pair and other media. The Lonworks network employs a connectionless domain-wide broadcast topology with loop-free learning routers and repeaters. Echelon have also designed the Neuron Chip, which builds in much of the Lonworks protocols and addressing, providing for economic development and a benchmark for the standard. This brings down cost and benefits installation.

While most of the Lonworks protocol is public and open, layers 3 to 7 of the standard are still closed and proprietary [8], along with Lonworks wiring system. Although developers have extended its application field, Lonworks still has issues with flexibility and scalability, which make other applications difficult to develop.

6.1.4   Universal Power Line Bus

Universal Power Line Bus (UPB) [9] is a mature, open low-rate PLC standard developed by PCS Power Line Systems for implementation on general-purpose microcontrollers. UPB is faster than other PLC based protocols and more reliable, however, its cost is a prohibiting factor as it requires manual installation and requires the configuration of dedicated devices and software making its wide adoption unlikely.

6.1.5   Homeplug

Homeplug [10] is another mature PLC protocol, initially developed to extend and adapt bus Ethernet networks and their devices while providing easy installation, reliability and high transmission speed that could reach up to 14 Mbps. There are many variants of Homeplug that can be used for different types of home automation and control applications, including: Homeplug AV' for Audio Visual (AV), 'Homeplug Command & Control (C&C)' for control of Heating Ventilation and Air Conditioning (HVAC), lighting and security and 'Homeplug Access BPL' for Broadband

Power line (BPL) and recently Homeplug Green PHY (GP)'. Homeplug C&C and GP are aimed at appliance control and Smart Grid and Smart Energy applications.

Homeplug's use of power line, its stability and its universal standardisation make the technology a strong candidate for ubiquitous deployment, but devices are large form factor and integration with small and dispersed sensors has not been a common feature. As such, there is not yet a clear catalyst for widespread adoption.

6.1.6 KNX

KNX [11, 12] is a mature and open but proprietary Home and Building Control standard owned by an industry alliance called the KNX Association. The standard may be obtained either through paid membership of the association or for a separate fee. KNX has gained a number of standards approvals including International Standard ISO/IEC 14543–3.

KNX power line operates in the 90–125 kHz band and at a bit rate of 1200 bps. Ease of installation is a key feature with KNX providing a single tool which supports all implementations and all user levels. KNX is mainly used for lighting and HVAC in commercial buildings rather than residences and mainly in its home country of Germany [13].

KNX shows some interest in Energy Management and Smart Grid, but its involvement in practice has been limited compared to Homeplug, Insteon and Lonworks. Moreover, the restricted openness of the specification and the high cost of membership and certification have been the main prohibiting factors to wide adoption. KNX support for multiple media is a strong asset, but it is also let down by a lack of security in KNX RF.

6.2 Ethernet

Ethernet is the standard of choice for wired local area data networks and it could easily be used to provide high quality communication for building appliance monitoring and control where wiring already exists or can be deployed. Ethernet is a mature protocol and is globally standardized by IEEE 802.3 [14] which defines the physical (PHY) and data-link (MAC) layers of a wired Local Area Network. The MAC layer protocol operates Carrier Sense Multiple Access with Collision Detection and specifies both half and full duplex operation. The PHY layer is run over coaxial, twisted pair, or fibre optic mediums and operation can range anywhere between 1 Mbs and 10 Gbps, though 100 Mbs or 1 Gbs is now common. Ethernet is a shared medium and relies heavily on upper layer protocols for security, though the risk is lessened by switched Ethernet. Also, although gateways and adapters exist to extend Ethernet to other media such as WiFi, ADSL, PLC and RF, the technology does not have an inherent affinity with other media and form factor is still not small enough for ubiquitous environmental sensing.

On the positive side, a number of relevant appliance communication standards that have been proposed to support Ethernet. For example, the KNX protocol for building control includes Ethernet amongst its communication media, and Home-plug was developed to extend or adapt an Ethernet network over power line. Ethernet also provides excellent support for common transport such as TCP/IP due to its long use in data networking and the terminals, though perhaps not the cabling is very cost effective.

Although a very popular standard for local area communication, Ethernet is less popular for appliance and energy management due to the high installation effort and cost of introducing new wiring, particularly in homes. These limitations may discourage its ubiquitous deployment within homes to connect appliances and electrical devices.

6.3   Wireless Communication

Since the start of the new millennia, wireless communication technologies have seen a sudden and sustained growth and in the last 2–5 years we have begun to see the emergence of the kind of technology that might support energy management through widespread environmental sensing. This has come through the inception and advancement of low-power, low-cost Radio Frequency (RF) wireless communications with smaller form factor, greater sensing density, and longer functional lifetime.

6.3.1   Wi-Fi

IEEE 802.11 was among the first RF standards for what are commonly known as Wireless Local Area Networks. Wi-Fi is the certification for WLAN IEEE 802.11 standard compliant devices by the Wi-Fi Alliance. The standard defines the characteristics of the PHY layer, including frequency and modulation methods, and the MAC layer using CSMA/CA with a binary exponential back off algorithm. The standard is an extension to the Ethernet protocol and often provides an access point to a variety of networks and media, providing wireless mobility and ease of installation while continuing to support common networking and transport protocols like TCP/IP.

The early (a/b) standards supported speeds of 1 Mbps at up to 100 m in a 20 MHz channel while the latest amendments (n) [15] support a 20 MHz channel and a 40 MHz channel with a data rate of up to 150 Mbps, and up to twice the range of the legacy standard.

Security is provided through the Wi-Fi Protected Access (WPA) specification which is based on Extensible Authentication Protocol (EAP) and Advanced Encryption Standard (AES), with the improved WPA2 providing CCMP, a new AES-based encryption mechanism.

However, while it is now ubiquitous in home data networks, Wi-Fi is not popular in the field of home automation as it has not been design for home energy management, particularly environmental sensing, as it is considered too power hungry and in some cases requires the installation of large components. However, a large number of amendments exist to the initial standard and the technology is now very mature, with low cost commodity status. These amendments were more recently rolled up into a new base standard, 802.11–2007 [16]. These amendments provide for a range of frequencies including 2.4, 3.7 and 5 GHz and a range of modulation methods including DSSS, OFDM and FHSS. However, the power consumption of these recent standards still exceeds that of Bluetooth (IEEE 802.15.1) Zigbee (802.15.4), and other Wireless Personal Area Networks (WPANs) standards.

6.3.2   IEEE 802.15.4 and Zigbee

In contrast to IEEE 802.11, IEEE 802.15.4 [17] is the standard for low power and low rate communications in Wireless Personal Area Networks (WPAN). At the physical layer, 802.15.4 specifies three different bands from amongst the available Industrial Scientific & Medical (ISM) frequencies for various locations and data rates, 868–868.6 MHz (1 channel, 20 kb/s), 902–928 MHz (10 channels, 40 kb/s) and 2.40–2.48 GHz (16 channels, 250 kb/s), and more recently 779–787 MHz (802.15.4c) and 950–956 MHz (802.15.4d). Different modulations are available for carrying data in the different bands, including DSSS and a choice of PSK types. Interfacing with Wi-Fi or Ethernet is common, but there is little affinity with other media such as PLC, IR or VLC.

At the MAC layer, devices can be full or reduced function (FFD or RFD), and each network segment has a single FFD co-coordinator node, responsible for management of the network segment (or Personal Area Network, PAN). FFDs may communicate peer to peer, but RFD may only communicate with the coordinator. Also, in IEEE 802.15.4, a network segment may operate in two modes: beacon mode, or non-beacon mode. If beaconing mode is enabled, the coordinator node will periodically send out a beacon with detailed information about the network segment and possibly guaranteed time-slots for certain nodes. In the time between the beacons nodes in the network may send data controlled by CSMA-CA. Non-beaconed networks simply send data using CSMA-CA without beaconing. However, while some security services are provided for upper layers by the MAC, extended security is expected to be implemented at higher layers.

Although IEEE 802.15.4 is mature wireless communication technology, it is not expected to be used for full end to end communication. Instead, it offers its lower layers functionality: physical layer, and MAC layer to higher level protocols. Among the high level protocols that use IEEE 802.15.4 as an underlying communication platform is ZigBee [18, 19, 20]. ZigBee is a self-organizing mesh open wireless networking standard ratified by the ZigBee Industry Alliance in 2004 [21]. The standard is optimized for low rate, low power, and medium range (10 m–75 m) communication and aims to provide devices with long functional life-time.

In ZigBee, the network layer should define the stack profile and the network rules, and implements discovery, addressing, routing and maintenance functionality. The stack then specifies the network rules such as level of security, timeout, sizes, maximum routers, children and depth of the network. To this end, ZigBee also specifies three ZigBee devices above 802.15.4, the controller, the router, and end devices. There can be only one coordinator device per network which initiates the network, and selects the network Id, stack profile and RF channel. At the MAC layer it is the Zigbee coordinator that acts as the 802.15.4 PAN coordinator. Routers then join this coordinator network, and other routers joining those routers, forming the mesh network. Once the network is formed, the coordinator also acts as a router and can perform other applications as required.

An end device joining the network will find and connect to a router or the co-ordinator, it has no children and performs no routing functions. This is the lowest powered node on the network and relies on its router to wake it up when required. At the MAC layer the Zigbee end device provides the 802.15.4 RFD function.

The ZigBee Alliance specifies a range of their own application profiles, such as ZigBee Home Automation, ZigBee Smart Energy, and ZigBee Health Care which would make it a good technology to use. ZigBee is now a mature and adopted standard but mass adoption and economies of scale still elude the ZigBee standard, with similar markets and returns for stakeholders not currently forthcoming. As with other such standards, the cost of entry set by the Alliance is certainly a limiting factor in the level of adoption. Also, the ZigBee networking standard does not support a common transport, but there are signs that the ZigBee Alliance may be moving away from this or at least breaking the sole dependency on IEEE 802.15.4. Version 2.0 of ZigBee Alliance Smart Energy Profile, whose requirements are specified in [21], is medium independent, using an IPv6-based common network and transport layer; implementations based on IEEE 802.15.4 will use the IETF 6LoWPAN adaptation layer.

6.3.3   Radio Frequency for Consumer Electronics

Radio Frequency for Consumer Electronics (RF4CE) [22] is an RF technology standard designed specifically to address the limitations of Infra-Red technology such as: field of vision, line of sight, one way communication and interference from other light sources such as large HD televisions. Like ZigBee, RF4CE is built on top of IEEE 802.15.4 PHY/MAC and operates on the 2.4 GHz frequency. The standard specifies simple networking and public application profiles, such as Consumer Electronics Remote Control (CERC), which interface to user applications. The public application profiles may include vendor specific extensions. In RF4CE, the network is an asymmetric star topology made up of two types of nodes, controllers and targets, though controllers need not be remote controls and could even be appliances.

Installation effort for RF4CE is relatively low, targets have the authority to check channel suitability, and for any pre-existing personal area networks (PANs), and then start a new PAN. Multiple controllers can join networks by pairing with targets and targets can communicate with other targets and join PANs to form a RF4CE network. Communication is protected using 128-bit cryptographic key pairs and services for confidentiality, authenticity and replay protection are included. Power management can be achieved via a range of configurations for enabling or disabling power saving mode, including operating a receiver duty cycle where the sender targets the active period. Channel agility means, after creating their PAN, target devices can switch channels if conditions dictate, and controllers will try other channels until communication is re-established.

While RF4CE's primary focus is an RF remote control replacement, it could equally be used to support many other applications, such as communication with small form factor automated controllers or sensors in an energy management system. In March 2009 the RF4CE Consortium teamed with the ZigBee Alliance to provide a new standard for RF remote control [22]. As with ZigBee, cost of license and certification still affects demand and widespread adoption and economies of scale are not yet fully realized.

Through RF4CE and its marriage with Consumer Electronics companies, the critical mass required, for wide adoption, and almost ubiquitous deployment for home automation including energy management, could be finally realized. Providing CERC and other profiles with two way RF communications opens the way to a range of new applications to attract consumers and manufacturers. Although RF4CE does not support energy management applications directly, there is support for additional profiles over the same network layer and with ZigBee's 'Smart Energy' profile there may be room for dual function. However, communication across the Internet using common transport would still require some kind of gateway or tunneling function.

6.3.4 Bluetooth

Bluetooth was the standard proposed by the Bluetooth Special Interest for a short range standard for industrial, home and office use. It used the 2.4 GHz ISM band and achieves a range of 10 m and a data rate up to 1–3 Mbps according to the original version.

The Bluetooth Transport Protocol Baseband layer defines a Piconet topology. Piconet recognizes a slave role and a mater role. Piconets are not formed through any central control, but are started by a master and can include slaves which register with the master. The master allocates addresses, unique within the Piconet, for active slaves, which may not exceed seven. Moreover, nodes may have dual role and can be master in one Piconet and slave in another forming a Scatternet. Piconet implementations have the potential to be self-installing and self-organizing.

In contrast to the above protocols that have struggled to find mass deployment, Bluetooth has been adopted for use on a verity of devices, including: mobile phones, PC USB dongles, remote printers, keyboard or mouse peripherals, game console controllers, etc. Bluetooth also has a range of supported protocols including TCP/

IP, making it a strong candidate for integration into a common transport. Bluetooth provides no direct support however for energy management applications. However, despite the possibility for small form factor, the inherent power demands may preclude the use of Bluetooth for, low power, long lifetime ubiquitous sensing nodes. For example, Bluetooth has a higher power cost to 802.15.4 and Zigbee, though a significantly higher data rate.

More recently, a more energy efficient version of the standard has been proposed by the Bluetooth group, called Bluetooth Low Energy' (BLE) [23, 24]. BLE uses a simplification of the Classic Bluetooth protocol with relaxed RF requirements, shorter packets and greater optimization of power when not transmitting. The Bluetooth Low Energy draft published in December 2009 includes Home Automation and RF remote control among its applications and its interoperability with the widely adopted Classic Bluetooth therefore make it a strong contender.

With the introduction of BLE, there are now two Bluetooth device types, single-mode devices which support BLE only and cannot communicate with a Classic Bluetooth device; and dual-mode 'Classic Bluetooth' devices which can also support BLE communication. The current trend is that dual-mode devices will replace all legacy Bluetooth devices over the next few years. Dual-mode devices are expected to bear much of the configuration load and provide Internet connectivity on behalf of low powered single-mode devices.

BLE uses the 2.4 GHz RF Band and an ultra-low duty-cycle transceivers, using advanced sniff-sub-rating, which allows a host to be woken by a controller for sending and so allows at least a year between single button battery cell recharge. BLE Connection setup and tear-down is also faster, giving low latency and speeds of up to 1 Mbps at ranges of 100 m. Accuracy and security are provided through strong 24 bit CRC and Full AES-128 encryption with CCM.

In addition, this new technology shares much of the original Bluetooth's radio characteristics and functionality and therefore mitigates cost and promotes manufacturer migration to dual-mode devices with the incentive that a whole range of new applications are enabled by connecting new low energy devices and accessories with an established Classic Bluetooth 1 billion unit base.

BLE is expected to see wide adoption as applications that connect small form factor sensing with personal mobile devices, such as phones, could see the technology quickly gain critical mass driving down cost and promoting near ubiquitous deployment. In addition to Home automation, Health Care and consumer electronics remote control are also some of the suggested applications for Bluetooth low energy.

6.3.5 *Visible Light Communication*

Visible Light Communication (VLC) [25] uses light as a medium to send data for short distance communication. Although the transmission range of VLC is relatively short (400–700 nm), it exhibits certain features that make it suitable for the set of ubiquitous communication desired in home energy management.

In addition to low power and high bandwidth, VLC is with communication that has many other attractive properties such as low noise, improved security and low cost. VLC is therefore considered as a good wireless communication and is more environmental friendly as it helps to reduce the exponential proliferation of Radio Frequency (RF) communication which is increasingly raising health and safety concerns as well as regulation conflict issues.

Recently, the light-emitting diode (LED) has undergone a serious evolution, with increased intensity, efficiency, control and greatly reduced cost and form factor. All of these factors make LED a genuine candidate as the ideal technology to implement the vision of easy installable and low cost ubiquitous VLC communication networks.

The IEEE 802.15 (WPAN) task group has recently released its 802.15.7 [26] standard that specifies the functionality of PHY and MAC standards for Visible Light Communications. In addition, there is a strong potential for affinity between VLC and power line communication which increases the chances for appliances, sensors, meters and other smart grid elements to ultimately interact together. However, though there is already strong industry collaboration and a number of proposed applications, there is yet no clear catalyst for widespread adoption.

6.4   IP Protocol

As we have seen throughout this chapter, IP technology will play major role in the realization of the HANs that will support the HEMS control operations by providing a unifying upper layer protocol and common communication medium. By utilizing IP technology in much the same way that it drove the development of the Internet, smart grid communications can exploit the vast range of features and services developed for these networks for its own purposes This includes support for differentiated quality of service classes, enhanced security, multicast operation, and robust and resilient routing while supporting a wide range of link layers, thus forming a comprehensive and highly flexible end-to-end architecture.

There are a wide range of IP-based technologies that have been designed, both to support private networks and in the public Internet, over the past three decades have would be highly valuable in the home networking environment so it makes a lot of sense to capitalize on these where possible. These IP technologies have been developed to build highly reliable and secure networks, which is fundamental to our requirements, and to support the demanding service characteristics for a variety of applications, while still supporting a wide range of link layers, thus forming a comprehensive and highly flexible end-to-end architecture. As such, it would be possible to build IP-based home networks for the smart grid that utilize a wide range of PLC, data, or wireless technologies as necessary whilst still expecting them to interact seamlessly.

The newest version of the IP protocol, IPv6, has been designed to meet the addressing demands of modern densely-deployed networks and, more recently, new

IP protocol suites have been proposed specifically for resources-constrained and low power transmission devices. For instance, new IPv6-based protocols have been defined in the Internet Engineering Task Force (IETF) to provide efficient header compression over low-speed link layers, a mechanism to support multicast, and new routing protocols that have been defined for large scale networks made of "low power and lossy" networks (called RPL) [27]. Moreover, the potential of this area is now well understood and further IP protocols and extensions are already being developed to adapt it to this scenario.

As such, while the HEMS will use a variety of networking technologies across the HAN as identified previously, we can expect that IP will provide the common medium for them to interact, just has it does with the Internet. However, the suit-ability of a particular IP protocol suite for the HEMS should be analyzed further against the requirement of the control application and the scale and scope of the HAN to identify which is more suitable to support the specific scenario. As the HAN that provides the communication infrastructure of the HEMS will, in reality, be a heterogeneous network with physical layer and link layers characteristics that vary from one part to the other, IP will be fundamental to allow these separate ele-ments to work together With this in place, it is easy to see how IP-based concepts and technologies could be used to build functional systems. For example, one could easily imagine a HEMS build around existing distributed systems, or web technol-ogy, providing interfaces via REST or SOAP, and passing messages in common formats such as XML.

6.5   Conclusion and Open Issues

We are beginning to see some exciting developments in communication and sensing technologies. Improved implementations of PLC and the emergence of new stan-dards such as Homeplug, Bluetooth Low Energy, 6LowPAN, and IEEE 802.15.7 Visible Light Communication (VLC) are providing greater bandwidth, reductions in message size and power, and the provision of two way communication sensing and on-board computation using harvested energy. As such, all of these technolo-gies are viable solutions to connect the different home appliances, controllers and intelligent sensors to the energy management platform.

However, given the diverse nature and location of elements across the energy management systems inside the premises, there is a clear need for a multi-technol-ogy solution. Hence, the home area network that will eventually emerge from the interconnection of all these entities will be using communication technologies as necessary and will thus be totally heterogeneous. It is therefore imperative to ad-dress a number of challenges that might arise from such cross-media communica-tion architectures.

While the level of penetration of each communication technology will vary from one deployment scenario to the other, certain research issues will need to be tackled in this research program:

- **Cross Layer Optimization**: It has been already proven that sharing information between different layers of a protocol stack can improve system performance, however; such an approach has only been considered in homogenous networks. The challenge here will be to study and devise novel cross-layer optimization approaches where a heterogeneous mixture of links and protocol stacks exist.
- **Interference Management and Routing Strategies**: As the number of appliances, controllers and intelligent sensors using the RF spectrum will increase in the HAN, some regions of the RF space are expected to become congested. The challenge will be to consider interference aware routing strategies that exploit knowledge of the physical radio channels across the network to route signals to avoid existing congestion or creating new congested regions. By understanding the properties of the radio channels, the network QoS can be managed more effectively and the channels used more efficiently (e.g. low rate data without tight latency requirements can be sent through poor channels with highly redundant codes).
- **Internetworking for Heterogeneous Wireless/Wireline Networks**: Different communication technologies such as PLC, Wi-Fi, or ZigBee have significantly different capabilities in terms of supported data rates and transmission/reception/control functionalities. However, the end-to-end performance requirements of the energy control application cannot be provided nor guaranteed without an efficient architecture to ensure proper internetworking functions for the different access technologies. For certain applications, a reasonable level of convergence for the access technologies is at the network layer (IP), so that the details and differences at the PHY and MAC layer for the access technologies can be "masked" out. On the other hand, while considering the performance impacts of the specific PHY and MAC protocols, other internetworking functions such as transport protocols and security among different access networks as well as between wireless access and wired networks needs to be defined and evaluated.

References

1. S. Galli, A. Scaglione, and Z. Wang, *For the Grid and Through the Grid: the Role of Power Line Communications in the Smart Grid*, in Proceedings of the IEEE, vol. 99, n. 6: pp. 998–1027, 2011.
2. Y. Son, et al., *Home Energy Management System Based on Power Line Communication*, IEEE Transactions on Consumer Electronics, vol. 56, n. 3: pp. 1380–1386, 2010.
3. D. Trichakis, et al., *Power Line Network Automation Over IP*, in IEEE International Conference on Telecommunications and Multimedia (TEMU), 2012.
4. D. Rye, *The X-10 POWERHOUSE Power Line Interface Model# PL513 and two-way power line interface model# TW523*, Technical Note, Revision 2, 2013.
5. P. Darbee, *INSTEON: The Details*, in Smarthome Technology, pp. 1–64, 2005.
6. M. Neugebauer, et al., *Automated Modeling of LonWorks Building Automation Networks*, in 2004 IEEE International Workshop on Factory Communication Systems.
7. M.R. Tennefoss, *Technology Comparison: LONWORKS® Systems versus DeviceNet®*, Echelon Corporation, 1999.
8. P.J. Allen, et al., *Web Based Energy Information and Control Systems: Case Studies And Applications*, The Fairmont Press, Inc, 2005.

9. *UPB, Technology Description, 2007.*
10. *Home Plug Alliance, Home Plug Green PHY The Standard For In-Home Smart Grid Power-line Communications*, June, 2010.
11. H. Merz, et al., *Building Automation: Communication Systems with EIB/KNX, LON and BACnet*, Springer, 2009.
12. KNK Association, *KNX Standard*, 2013.
13. A. Kell, and P. Colebrook, *Open Systems for Homes and Buildings: Comparing Lonworks and KNX,* Watford, UK: i&i limited, 2004.
14. LAN/MAN Standards Committee, *IEEE Standard 802.3–2008*, IEEE Computer Society, 2008.
15. Y. J. Wen, and A.M. Agogino, *Wireless Networked Lighting Systems for Optimizing Energy Savings and User Satisfaction*, in IEEE Wireless Hive Networks Conference, 2008.
16. J.S. Sandhu, A.M. Agogino, and A.K. Agogino, *Wireless Sensor Networks for Commercial Lighting Control: Decision Making With Multi-Agent Systems*, in AAAI Workshop on Sensor Networks, 2004.
17. E. Karapistoli, et al., *An Overview of the IEEE 802.15.4a Standard*, in IEEE Communications Magazine, vol. 48, n. 1: pp. 47–53, 2010.
18. ZigBee Alliance, *ZigBee Specification*, 2008.
19. N.A. Somani, and Y. Patel, *Zigbee: A Low Power Wireless Technology for Industrial Application*, in International Journal of Control Theory and Computer Modelling (IJCTCM) vol. 2, 2012.
20. ZigBee Alliance, *Latest ZigBee Specification Including the PRO Feature Set*, 2005.
21. ZigBee Alliance, and Home Plug Alliance, *Smart Energy Profile 2.0 Technical Requirements Document*, April 2010.
22. ZigBee Alliance, Understanding ZigBee RF4CE, 2013: https://docs.zigbee.org/zigbee-docs/dcn/09-5231.PDF.
23. R. Heydon, N. Hunn, Bluetooth Low Energy, 2010: https://www.bluetooth. org/DocMan/handlers/DownloadDoc.ashx.
24. C. Gomez, J. Oller, and J. Paradells, *Overview and Evaluation of Bluetooth Low Energy: An Emerging Low-Power Wireless Technology,* in Sensors, vol. 12, n. 9: pp. 11734–11753, 2012.
25. S. Schmid, et al., *Visible Light Communication: Combining Illumination and Communication*, in 2014 ACM SIGGRAPH Emerging Technologies. ACM, 2014.
26. S. Rajagopal, R.D. Roberts, and S.K. Lim, *IEEE 802.15. 7 Visible Light Communication: Modulation Schemes and Dimming Support*, in IEEE Communications Magazine, vol. 50, n. 3: pp. 72–82, 2012.
27. E. Ancillotti, R. Bruno, and M. Conti, *The Role of the RPL Routing Protocol for Smart Grid Communications*, in IEEE Communications Magazine, vol. 51, n. 1: pp. 75–83, 2013.

Chapter 7
Towards a Unified Smart Grid ICT Infrastructure

Throughout the previous chapters we have outlined the key technologies that will characterize the development of the smart grid. Innovative new power engineering technologies will be required to introduce flexibility so the power distribution network can incorporate new distributed generation systems and optimize delivery in a more responsive and dynamic environment. Moreover, an entirely new plane is introduced in the form of smart metering feeding back from end users towards energy suppliers and network operators and ultimately provide an end-point for dynamic two way communication between producers, providers, and consumers. As we have seen, the provision for these new technologies necessitates the provision for ubiquitous data communication networks, far beyond what is currently in place, and this has been the main focus of this work. However, while we have seen how communications networks can be built around the technologies considered here, we have yet to show how these systems could be composed together and what functionality will be necessary to govern the interaction between these disparate elements.

The final chapter of this book will therefore explore the topic of a unifying smart grid architectural design in more detail and describe the core elements that will be necessary to achieve this. The first aspect that we will consider is the issue of data management beyond the aggregation network considered in chap. 3 and how this can be harnessed by the utilities. Thereafter, we will look at computing architectures and how the necessary storage and processing resources can be provisioned in the smart grid. Finally, we will explore the end-to-end architecture and outline how the various elements can interact.

7.1 Architectural Design and Data Aggregation

Given the scale and scope of the systems being considered, with millions of potential end customers, the architecture for a smart grid computing and communication system must be carefully designed. In particular, the development of a distributed data aggregation model will be critical to make the system scalable and responsive.

© The Author(s) 2014   83
F. Bouhafs et al., *Communication Challenges and Solutions in the Smart Grid*,
SpringerBriefs in Computer Science, DOI 10.1007/978-1-4939-2184-3_7

A key decision here will be in choosing between a semi-centralized or wholly distributed model whereby the data is managed either by a small set of entities or performed by many entities respectively. This involves calculating the trade-off between simple, unified control on one hand and more localized, responsive but complex control on the other.

7.1.1   Smart Grid Data Aggregation

One of the key challenges in the smart grid will be properly managing the streams of data generated by the introduction of smart meters such that they can be used for effective grid management. Since smart meters are still being introduced in many areas, the standards for the types and frequency of data sent from these devices is still to be finalized. However, given that there are upwards of 26 million households in the UK, and over 100 million in the USA, even relatively conservative numbers here would cause a significant issue. Moreover, while periodic reporting will be sufficient for basic metering purposes, if a utility is going to use smart meter data to monitor the state of the distribution network it will need far more active data to be available, perhaps live streaming at busy periods and/or if the home is generating its own electricity supply. As such, the scalability issues introduced here necessitate data management approaches to be introduced.

In previous chapters, we described the concept of using local data collector nodes as a first point of contact in the AMI system, particularly in congested urban environments. This would also offer a first point of control for data management by the utility companies. By exploiting data sampling techniques it is possible to reduce the data rate of the readings generated by all of the connected smart meters and then aggregate them together to minimize the amount of data send into the network. This has the advantage of providing an accurate reading local to a specific neighborhood while limiting load on the network core.

This same technique could also be utilized in the core of the network where the data sent by all the various collection points are sent for further processing. It is possible to envision a hierarchy of nodes in such an AMI network whereby groups of collection points are formed, perhaps around the micro-grids discussed in chap. 2, such that information relevant to a particular area is made available within that scope on demand but also aggregated and provided to higher level entities enforcing the scalability of the system. At the top level, the grid operator will then have access to data from the entire grid but at a reduced granularity of detail unless specifically required, as shown in Fig. 7.1.

An alternative approach would of course be to remove the complexity of implementing a multi-level architecture and simply provision a single, centralized system capable of accepting data from all the smart meters within the grid and processing and reporting to the grid provider as required. Of course, any such discussion should take into account the issues of system resilience and data security. A more distributed approach might be robust but is also potentially more open to attack or data loss.

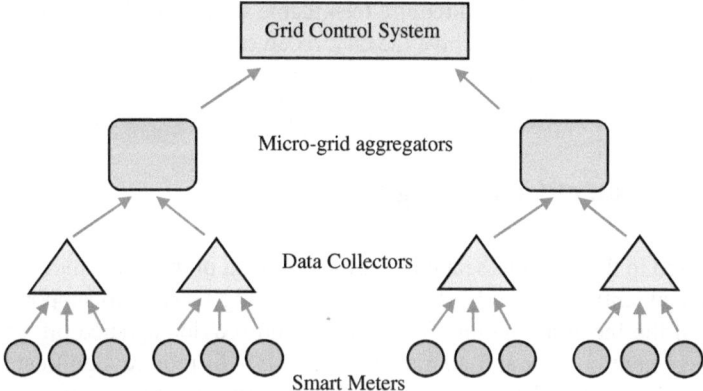

Fig. 7.1 Data aggregation techniques in the network

7.1.2 Interface to the Grid Control System

Regardless of the architecture of the data aggregation system, the second key aspect of this part of the smart grid will be in defining the interface between this system and the grid operator/ utility control system. For security reasons, any system that is accessible from the public will be kept entirely separate from the grid (SCADA) control systems to eliminate the potential for accidental of malicious impact on the operation of the grid. As such, the interface that exists between these two systems will need to be carefully designed in order to provide access to the data but minimize the risks involved.

First it is necessary to define the nature of the interface between the data aggregation network and the grid control systems and the types of data to be sent and received. In this case, the interaction will primarily be one-way in that data will be requested by the control system and provided by the aggregation system. Alternatively, some of the data might be constantly streamed to the control system and the control system might occasionally issue configuration commands to the aggregation system. This minimizes the potential for the data plane to affect the control plane, except as a result of the data it provides.

In practice, it is possible that the control system is also distributed such that the services that interact with the data aggregation service are isolated from the grid operator. As such, the control system must implement services that gather data from the aggregation network and perform some processing to determine the impact on the current state of generation and the distribution network. In practice, this these commands will flow via the grid controller to maintain the manageability and security of the model.

The control system will therefore not be directly responsible for managing the data collected by the smart meters, as it is with sensors deployed within the distribution network, and so a new entity is needed to provide this functionality. This entity will have sufficient capacity to receive all the data collected by smart meters within a specific grid network and will require formidable storage and processing capabilities in order to marshal and archive the data received and respond to queries issued by the grid control system.

7.2   Provision of Computing Power

As discussed in the previous section, a key component of the new smart grid system will be a new entity to provide 'back-end' processing services to support dynamic analysis of the incoming monitoring statistics and provide updated information to the management entities and administrators. This will require significant storage capacities and the provision of on-demand computing facilities to deal with large volumes of data. A natural candidate to fill this role is some form of cloud computing whereby resources can be allocated on-demand for data analysis without the need for significant initial outlay on equipment. However, there are still many issues related to cloud computing in this context that need to be resolved and perhaps the most significant of these is data protection and security, as strict guarantees will need to be given in this respect. This section will explore the requirements for this service and how cloud computing could fulfill this role.

7.2.1   Use of Cloud Computing for Data Management

Cloud computing platforms are ideally suited to provide this type of data storage and processing service because of the way they have been developed and can be provisioned. While a full discussion of cloud computing is beyond the scope of this publication, it is necessary to first outline the basic paradigm employed and its capabilities and potential drawbacks. Cloud computing emerged as a result of the convergence of multiple fields of computer science including; utility (grid) computing, large-scale distributed systems, and service virtualization, among others [1, 2]. In summary, a cloud service provider deploys one or more data center type computing centers (essentially a server farm) with the intention of providing a portion of these resources on-demand to a range of users. The user then requests a portion of the total resource based on their requirements and the service provider provisions an appropriate set of resources (through virtualization) in response, as shown in Fig. 7.2. The service provider is responsible for managing the underlying infrastructure and charges the user a fee for the services consumed in a particular period. The advantage here is that the user does not have to incur the cost of deploying and managing the equipment, while the service provider can exploit multi-tenant virtualization paradigms to maximize resource utilization and gain a profit through economies of scale. Moreover, the user can interact with their cloud resources regardless of their location, as long as they have an Internet connection and utilize the provider-defined APIs. Also, due to the scale of modern cloud computing platforms,

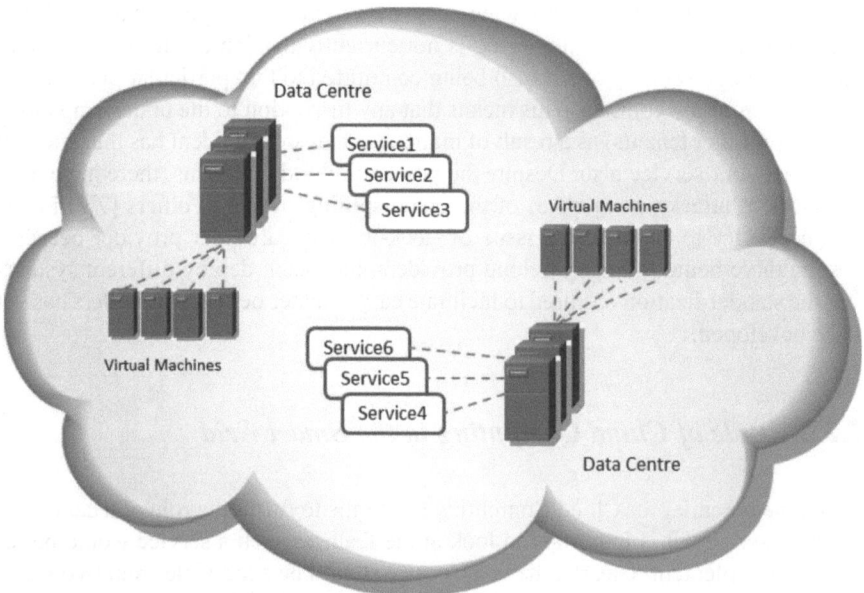

Fig. 7.2 Cloud computing model

the provider can automatically and dynamically enforce data/service replication to avoid loss or service disruptions [3].

It is therefore clear that Cloud computing has a great number of potential advantages that could be utilized for the data management service envisioned for the smart grid. The provider can implement this service, either directly or via a third party, on a pre-existing cloud platform and not incur the problems that such a large project could potentially involve. This means that costs will be minimized to the price of the resources used, and the lead-time required will be dramatically reduced. Of course, the software system that runs on the cloud service will still need to be developed regardless of the underlying system but, depending on the type of cloud computing used, the development environment may also be provided by the cloud provider. Moreover, one key advantage of cloud computing is that computing resources can be provisioned dynamically to services such that an initial deployment can start relatively small and grow over time, or additional resources can be provisioned on-demand to deal with short-term spikes or surges [4].

Of course, any such system is not without its shortcomings and Cloud Computing has several issues that should be carefully considered prior to deployment [5, 6]. Fundamentally, the first issue is that by employing computing resources from a third party one automatically gives up some control of the service to the cloud service provider. This implies that a certain degree of trust must be placed in them to maintain the confidentially of the user services and data and provide the features expected in terms of service availability, data replication and encryption, etc. as agreed in the contract or Service Level Agreement (SLA). However, many cloud providers are somewhat reticent about allowing customers to verify the performance of their platforms and third-party monitoring is not currently available. Moreover, by employing a cloud service, one is also being committed to that particular platform for the duration of the contract. This means that any disruption to the underlying cloud service (or other tenants) as a result of malicious attack or accident has the potential to impact your service also. Despite the power of cloud platforms, there have been instances of attacks on one part of the cloud spilling-over into others [7]. Finally, it is necessary to consider the issue of 'lock-in' to a particular provider because, despite there being a range of cloud providers, they each deploy different systems and the standardization required to facilitate easy transfer between providers has yet to be developed.

7.2.2   *Role of Cloud Computing in the Smart Grid*

Given the potential for Cloud Computing platforms to fulfill the role of a data management service, it is important to look at the features such a service would be required to implement. Clearly, the data service should be accessible from two sides, the smart meters via the aggregation network and the grid controller, and in general we can expect the cloud service to perform the role of collecting the data from the smart meters and organizing and storing it appropriately before processing it and reporting the findings back to the grid controller. The cloud service might also be expected to respond to specific requests issued by the grid controller and even issue service updates back to smart meters as part of the AMI. This section will explore these features in more detail.

In terms of data collection, one immediate advantage of cloud computing is that the server farms that make up the cloud infrastructure are typically heavily provisioned by their Internet Service Providers (ISPs) to cope with the immense traffic load, they also typically include peering to multiple Internet Exchanges through various ISPs in order to avoid overloading a specific path. As such, by load-balancing the incoming smart meter data, a cloud service could easily cope with the traffic demands. As such, the first task for the data management service would be to effectively marshal and store the data within the infrastructure. Typically the web service that receives the data will be connected to a database system where the data is stored. Again, the specifics of such a service are beyond the scope of this publication but they are designed to be sufficiently scalable for applications of this type, typically now using advanced SQL-based architectures or NoSQL approaches. Such databases will also be replicated and backed-up by default [8].

With the data in the system, the next task of the data management service will be to process the data and present it to the grid controller. Again, because the applications to manage smart meter and AMI infrastructures has yet to be defined, this is not standardized and no *de facto* approach exists. However, it is clear that cloud computing can provide the processing resources to support this. On a simple level, the service should provide a usage monitoring feature based on localities or

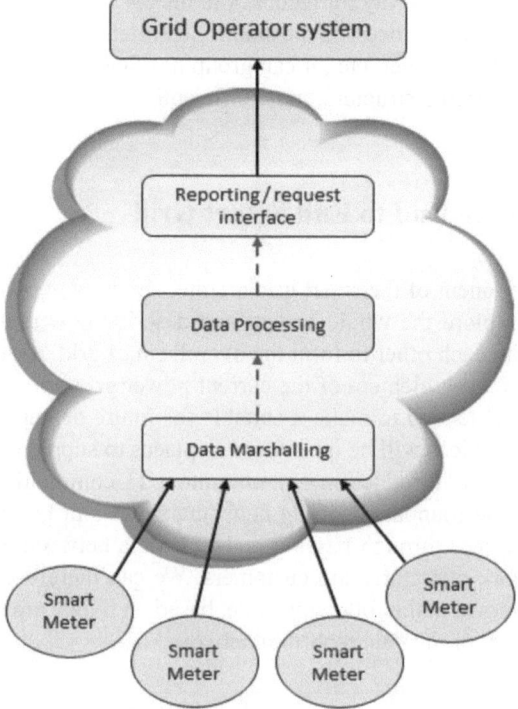

Fig. 7.3 Smart meter data management via the cloud

micro-grids of smart meters to identify or predict any peak loads in the grid. This will require the service to group nodes based on the locality and process the aggregate demand. Beyond this, the cloud service may be required to periodically provide more thorough analytical reports on the data which could be significantly more computationally expensive to perform. Such processes could easily be classed as Big Data problems and utilize MapReduce (or similar) applications to process the data efficiently [9]. As a result of these processing tasks, the data service should then report its findings to the grid controller service, as shown in Fig. 7.3.

Finally, the data aggregation service deployed on the cloud may also be required to implement services to process specific requests from the grid controller or pass notifications back to the smart meters, for example to inform them of changes in electricity pricing. Each of these aspects could utilize specific features of the cloud platform to simplify implementation. In terms of the first service, is it clear how this could take advantage of the highly dynamic nature of cloud computing to provision additional resources and deal with specific requests for subsets of the data. For example, as the archive of stored data increases, it should be possible to run more complex queries over the entire dataset to identify trends or predict future usage over monthly, annual, or longer periods. With regard to the second service, because the cloud acts as an endpoint for interaction with the smart meters, it is uniquely positioned to act as the gateway for feedback to the consumers. Moreover, since the service will already have processed the smart meter data and hold accurate data on their location, utility provider, etc., it can group notifications and process changes, for example to the pricing structure, more efficiently.

7.3   Building the End to End Smart Grid

With the last component of the smart grid architecture now in place, we are finally in a position to explore the whole system and describe how these numerous elements interact with each other to form the overall smart grid. As we have explored in this publication, each element of the current power grid will need to be significantly updated or extended to make it suitable for future demands on the network and entirely new services will be introduced in places to supplement existing functionality. Crucially the introduction of a ubiquitous IT communications infrastructure will provide the foundation of the new smart grid and facilitate many of the new enhanced services through passing real-time data between generator entities, distribution network operators, and customers. We can therefore identify the elements of the smart grid in the following table, based on if they are part of the energy distribution network or the data reporting network

Energy Distribution	Data Reporting
Power grid operator—control room	Homes (EMS, HAN, sensors)
Power plants	Smart meters
Transmission network	Data collector + NAN
Substations (intelligent controller)	Backhaul network
Distributed Generation	Data management service (Cloud)
Distribution network (sensors)	Power grid operator—control room
Consumers	

This section will draw together all the previous chapters, including the services introduced in sect. 2 above, to describe the operation of the new architecture and interoperation between the elements we have identified.

7.3.1   Power Station to Smart Meter

The first section will explore how the electricity generation and distribution network will be affected in the smart grid. In this case, most countries will already have a considerable existing infrastructure as outlined in chap. 1 with traditional, remote, energy generation stations connected to population centers via a transmission network. Once in the locality of consumers, the transmission network will connect to power substations and be delivered to homes and businesses via an extensive distribution network. As such, we expect that this infrastructure will remain largely in place, though still subject to ongoing modernization over time, and will be managed by the grid operator via a SCADA-based control system.

However, the nature of this control system is expected change dramatically in the smart grid with the introduction of distributed control based on the concept to self-contained micro-grids. This means that, while the grid operator will maintain overall control of the network, individual areas will be managed in fine-grained detail by local Intelligent Controllers located in the substations. This will allow micro-grids to balance their own network better and cooperate with neighboring Intelligent Controllers via a Multi-Agent System to import/export energy supply as needed. This will support the introduction of local generation much more effectively and will allow micro-grids to take into account the existence of storage elements and PHEVs to predict and plan for variations in demand and supply. This could also, for example, be supplemented with weather forecasting data to predict potential energy generation from renewable sources within the area.

The Intelligent Controllers will also provide an aggregation point for the ubiquitous deployment of sensors in the distribution network micro-grid to provide much more detailed information about the state of the network at any point in time and this is ultimately expected to be supplemented by information provided by consumer smart meters via the data management service and grid operator as described in sect. 2. The Intelligent Controller will also represent an enforcement point for the grid operator in the micro grid, acting as a repository to gather data about the state of the local grid and a control mechanism through which actuation commands can be passed.

7.3.2   Smart Meter to Grid Operator

In contrast to the energy distribution network, the usage monitoring and feedback aspect of the smart grid will in most cases need to be built from the ground up with no pre-existing elements in place. The first step here will be the deployment of smart meters in homes and businesses that can monitor usage and report back to the utility operators. They will also ultimately form the core of an Energy Management Systems (EMS) in buildings where smart appliances can monitor their usage dynamically and link this to the cost of the energy supplied. By networking these devices, along with local sensors, into an integrated HAN network, the EMS can control the consumption of energy within premises in a fine level of detail.

Beyond the smart meters, the utility will deploy an aggregation network to collect the meter reading and report them to the utility company. This could be via Data Collectors and NANs in densely populated urban areas as outlined in sect. 3 or directly to the operator otherwise and depending on the scenario, a range of communication technologies could be used to achieve this. Ultimately, this data will enter the Backhaul network until it reaches the Data management service, which is likely to be based on some form of Cloud Computing. The data management service

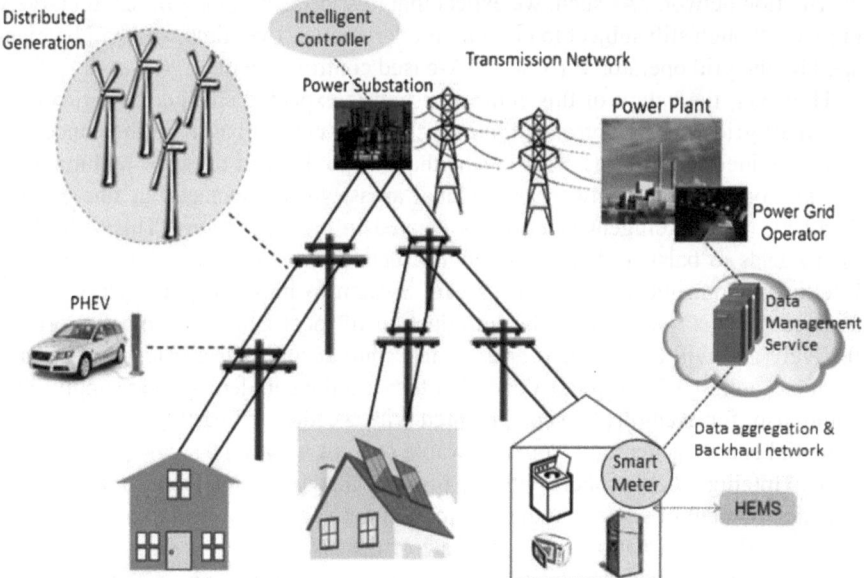

Fig. 7.4 Proposed smart grid architecture

will collect the meter readings from across the entire network and store it for further processing. From there, the data is analyzed and report back to the grip operator via the control room services.

Once the grid operator has received the data, it will perform its own analysis and apply any changes to the distribution grid as necessary. For example, metering data for a specific micro-grid might be periodically disseminated down to the Intelligent Controllers identified in sect. 3.1. It will also use the information for a range of other purposes such as billing and energy provisioning. Finally, the utility operator will also use this reporting network in reverse to issue updates on electricity pricing, etc., to the data aggregator which will in turn disseminate this information back down to the smart meters.

This two-way feedback mechanism, from the smart meter to the grid operator and vice versa, is a fundamental aspect of the smart grid and forms the core of the AMI and DR services described in chap. 3. This greater availability of information also allows for fine-tuning of the distribution network based on consumer demand and the adoption of more intelligent energy usage at the point of consumption, based on current network conditions. As such, we propose that it is this functionality, based on ubiquitous data networking, which will truly make the energy grid smart in the future and enable it to meet the demands of future users. Finally, we present every element of the smart grid architecture together in Fig. 7.4.

7.4   Conclusion and Open Issues

We have now broadly explored the elements and operations necessary to deploy the smart grid as it is currently envisioned. However, one thing that is clear at this stage is that the smart grid in its ultimate form is going to represent a very complex multi-provider system, several steps beyond what is currently deployed and so a wide range of challenges still exist. Beyond the technology-based issues discussed in this publication the broader aspects of ensuring that the smart grid, and particularly the data communications plane, meets the core requirements that have been identified raise a number of significant challenges. Moreover, while it is possible to identify the elements required for a smart grid, a range of more pragmatic issues arise once one considers how a system this expansive and complex can be deployed in a managed way. As such, this section will explore these communications challenges in more detail and suggest potential solutions where possible before looking more generically at the issue of deploying the smart grid.

7.4.1   A Simple, Scalable and Efficient System

As we have seen in previous chapters, for any system to be deployable over the entire network, it must be sufficiently simple and scalable for it to be seen as a worthwhile investment. For example, we have reviewed the issues of managing data aggregation from millions of isolated monitoring and metering devices and how specific measure must be used to ensure that a system is capable of dealing with the expected load. However, this is a highly complex issue and includes taking into consideration a range of other factors such as the cost of equipment in the core and edge, administrative and training overheads, development and maintenance costs, etc. This is because the system must be viewed in its entirety and cannot be viewed in isolation.

As such, one solution being considered in the context of the smart grid is to employ the services of third party providers who have the expertise and resources to manage specific aspects of the overall service. One natural example of this might be the cloud-based data management service that could be hosted by an existing cloud provider on a pay-as-you-go basis. Another such example would be the provision of dedicated 3G wireless links to connect isolated consumers, which could be sub-contracted to existing mobile network operators. However, while this solution has some obvious advantages, care must be taken to properly manage the interoperation of these separate services as the potential exists to introduce additional complexity and inefficiencies.

7.4.2   Secure, Robust and Reliable Communication

With the provision of a dedicated communication network that forms a critical aspect to the operation of the system, it is natural that this network must be made secure, robust and reliable. It has long been identified that such critical infrastructures must be protected and made resilient to defend against failures and attack, and mature network 'hardening' techniques can be employed in this case. In the event that the public Internet is used to interconnect customer equipment with the provider infrastructure it is obvious that strong encryption and authentication measure be adopted in the first case to ensure the security of data in transit. Moreover, energy providers may also consider adopting a trusted platform model to ensure the integrity of the entities in the system and help protect against attack or exploitations.

Moreover, it will be necessary to build redundancy into the communications system such that failures or attacks can be mitigated until remedial action is taken. Some good examples of this would be through the use of Cloud Computing mechanisms to ensure data and service replication and isolation, and the utilization of robust Multi-Agent Systems for micro-grid management. In the same way that failures are currently handled in Internet engineering, it should be possible to build sufficiently reliable systems that meet the requirements of the element in question. So, because the data management service is not mission critical, it can tolerate a certain degree of failure whereas the micro-grid management cannot and must be built to handle failures.

7.4.3   Smart Grid Deployment Pathways

The final issue that we address in this text is the problem of deploying a smart grid, such that the elements we discuss here are fully implemented. Of course, such a complex and multi-faceted system cannot and will not realistically be deployed in its entirety and in isolation. As we have already discussed, most if not all countries already have mature pre-existing energy grids and so any deployed must first take into careful consideration the state of this grid and how smart grid elements can be seamlessly deployed into it. This is, for example, why we propose a range of communications technologies in each context as no two grids will have exactly the same existing network, economic resources, geographic considerations, political backing, etc. As such, we can only make broad statements here to outline potential routes for adoption, and leave it to individual operators to consider the practical realities in each specific case.

Clearly therefore, the first point to make is that any smart grid deployment will proceed incrementally, with specific elements layered into the current infrastructure before the next is introduced and so-on. One example of this is the current drive towards introducing smart metering in many European and American power networks. Such deployments can largely be completed in isolation, with minimal impact, and represent a precursor to many other smart grid elements. Once this is completed (which is in itself no small undertaking), it opens the door for AMI and data aggregation and management services, and so on. Another current example is the gradual adoption of local renewable generation devices, though this is done on a far more individual basis.

Another potential solution to this issue is to adopt a modular approach whereby 'smart grid islands' can be deployed in the scope of a street, community, village, etc., and can be interconnected and scaled up from there. This approach utilizes the concept of building self-contained micro-grids including, for example, one or more DER generation devices, a set of homes in a specific region and an Intelligent Controller that can be deployed to manage them. Such an arrangement would, for example, be well-suited to isolated regions which a centralized grid operator typically struggles to supply effectively. This approach would also be useful as a 'pilot' or demonstrator deployment to test and evaluate the technologies used ahead of a broader roll-out.

However, regardless of the technologies eventually utilized, or the deployment model followed, it is clear that the smart grid is the future of energy generation and distribution and many of the features outlined here will, over time, form the foundation for how we receive energy.

References

1. M. Armbrust, et al. *Above the clouds: a Berkeley view of cloud computing*, UC Berkeley Technical Report, 2009.
2. NIST Definition of Cloud Computing v15, 2009, http://www.nist.gov/itl/cloud/upload/cloud-def-v15.pdf
3. C. Clark, K. Fraser, A. Hand, J. Hansen, E. Jul, C. Limpach, I. Pratt, and A. Warfield. *Live Migration of Virtual Machines*. In Proc of the Symposium on Networked Systems Design and Implementation, 2005.
4. Y. Song, Y. Sun, H. Wang, X. Song. *An adaptive resource flowing scheme amongst VMs in a VM-based utility computing*. 7th IEEE International Conference on Computer and Information Technology, CIT 2007, pp. 1053–1058.
5. D. Catteddu, G. Hogben. *Cloud Computing: Benifits, risks and recommendations for information security*. European Network and Information Security Agency (ENISA), 2009.
6. Cloud Security Alliance, "Top Threats to Cloud Computing", http://www.cloudsecurity-alliance.org/topthreats, March 2010.
7. D. C. Marinescu. "Cloud Computing; Theory and Practice." Morgan Kaufmann, 2013.
8. K. Grolinger, W. A. Higashino, A. Tiwari and M. A. Capretz, *Data management in cloud environments: NoSQL and NewSQL data stores*. Journal of Cloud Computing: Advances, Systems and Applications, 2013.
9. J. Dean, S. Ghemawat. *MapReduce: simplified data processing on large clusters*. In: Proc of OSDI, 2004.